Core Mathematics 1

Edexcel AS and A-level Modular Mathematics

Greg Attwood
Alistair Macpherson
Bronwen Moran
Joe Petran
Keith Pledger
Geoff Staley
Dave Wilkins

Contents

The highlighted sections will help your transition from GCSE to AS mathematics.

About this book

This book is designed to provide you with the best preparation possible for your Edexcel C1 unit examination:

- The LiveText CD-ROM in the back of the book contains even more resources to support you through the unit.
- A matching C1 revision guide is also available.

Finding your way around the book

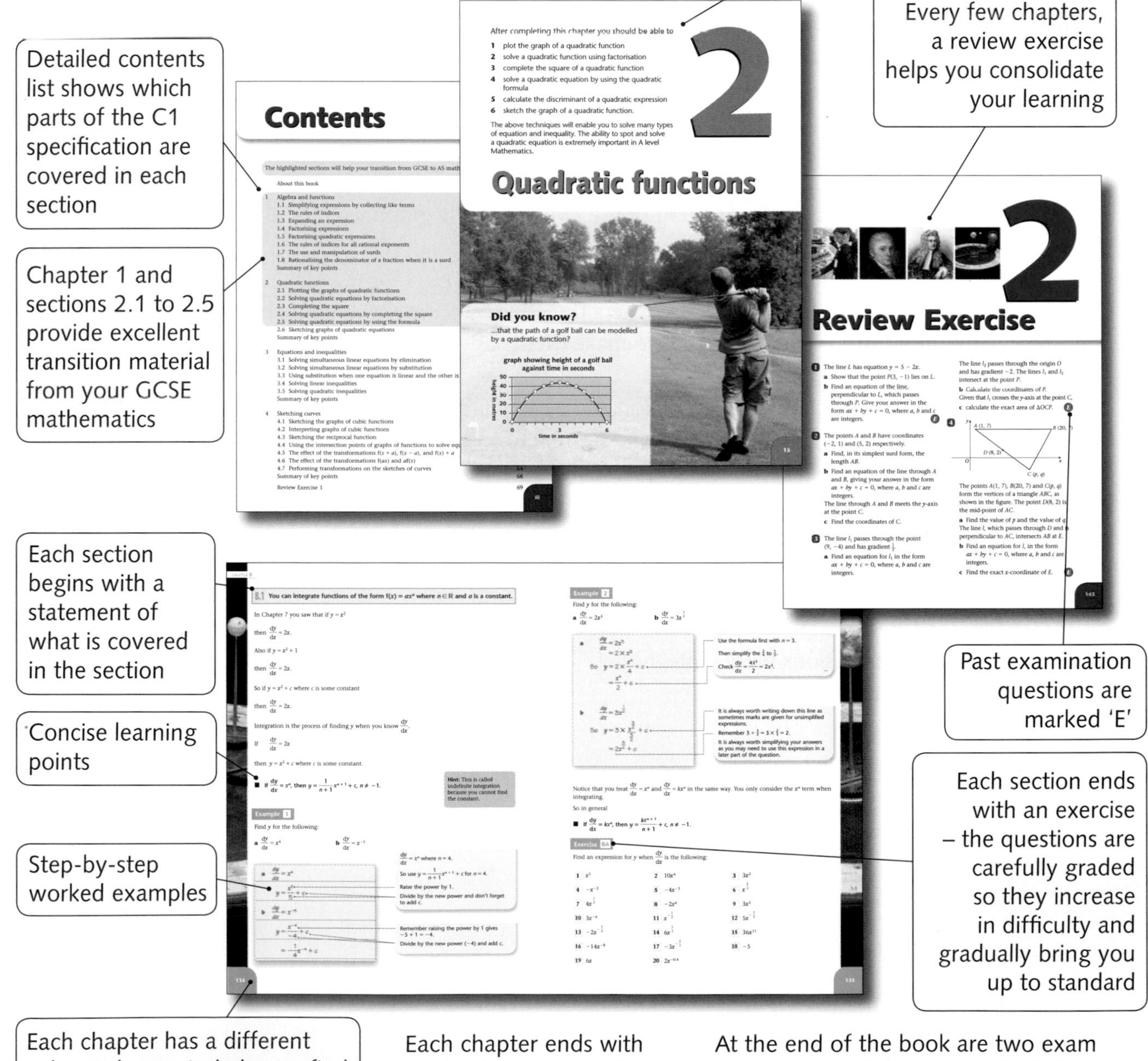

Each chapter ends with a mixed exercise and a summary of key points.

At the end of the book are two exam papers: a practice paper and a full examination-style paper.

LiveText software

The LiveText software gives you additional resources: Solutionbank and Exam café. Simply turn the pages of the electronic book to the page you need, and explore!

LiveText Core Mathematics 1

8.1 You can integrate functions of the form $f(x) = ax^n$ where $n \in \mathbb{R}$ and a is a constant.

In Chapter 7 you saw that if $y = x^2$

then $\frac{dy}{dx} = 2x$.

Also if $y = x^2 + 1$

then $\frac{dy}{dx} = 2x$.

So if $y = x^2 + c$ where c is some constant

then $\frac{dy}{dx} = 2x$.

Integration is the process of finding y when you know $\frac{dy}{dx}$.

If $\frac{dy}{dx} = 2x$

then $y = x^2 + c$ where c is some constant.

■ If $\frac{dy}{dx} = x^n$, then $y = \frac{1}{n+1}x^{n+1} + c$, $n \neq -1$.

Hint: This is called indefinite integration because you cannot find the constant.

Example 1

Find y for the following:

a $\frac{dy}{dx} = x^4$ b $\frac{dy}{dx} = x^{-5}$

a $\frac{dy}{dx} = x^4$

$y = \frac{x^5}{5} + c$

$\frac{dy}{dx} = x^n$ where $n = 4$.
So use $y = \frac{1}{n+1}x^{n+1} + c$ for $n = 4$.
Raise the power by 1.
Divide by the new power and don't forget to add c.

b $\frac{dy}{dx} = x^{-5}$

$y = \frac{x^{-4}}{-4} + c$

$= -\frac{1}{4}x^{-4} + c$

Remember raising the power by 1 gives $-5 + 1 = -4$.
Divide by the new power (-4) and add c.

134

Example 2

Find y for the following:

a $\frac{dy}{dx} = 2x^3$ b $\frac{dy}{dx} = 3x^{\frac{1}{2}}$

a $\frac{dy}{dx} = 2x^3$

$= 2 \times x^3$

So $y = 2 \times \frac{x^4}{4} + c$

$= \frac{x^4}{2} + c$

Use the formula first with $n = 3$.
Then simplify the $\frac{2}{4}$ to $\frac{1}{2}$.
Check $\frac{dy}{dx} = \frac{4x^3}{2} = 2x^3$.

b $\frac{dy}{dx} = 3x^{\frac{1}{2}}$

So $y = 3 \times \frac{x^{\frac{3}{2}}}{\frac{3}{2}} + c$

$= 2x^{\frac{3}{2}} + c$

It is always worth writing down this line as sometimes marks are given for unsimplified expressions.
Remember $3 \div \frac{3}{2} = 3 \times \frac{2}{3} = 2$.
It is always worth simplifying your answers as you may need to use this expression in a later part of the question.

Notice that you treat $\frac{dy}{dx} = x^n$ and $\frac{dy}{dx} = kx^n$ in the same way. You only consider the x^n term when integrating.

So in general

■ If $\frac{dy}{dx} = kx^n$, then $y = \frac{kx^{n+1}}{n+1} + c$, $n \neq -1$.

Exercise 8A

Find an expression for y when $\frac{dy}{dx}$ is the following:

1 x^3	2 $10x^4$	3 $3x^2$
4 $-x^{-2}$	5 $-4x^{-3}$	6 $x^{\frac{2}{3}}$
7 $4x^{\frac{1}{2}}$	8 $-2x^6$	9 $3x^5$
10 $3x^{-4}$	11 $x^{-\frac{1}{2}}$	12 $5x^{-\frac{3}{2}}$
13 $-2x^{-\frac{3}{2}}$	14 $6x^{\frac{1}{3}}$	15 $36x^{11}$
16 $-14x^{-8}$	17 $-3x^{-\frac{2}{3}}$	18 -5
19 $6x$	20 $2x^{-0.4}$	

135

Resources 6/7 of 9

Unique Exam café feature:

- Relax and prepare – revision planner; hints and tips; common mistakes
- Refresh your memory – revision checklist; language of the examination; glossary
- Get the result! – fully worked examination-style paper

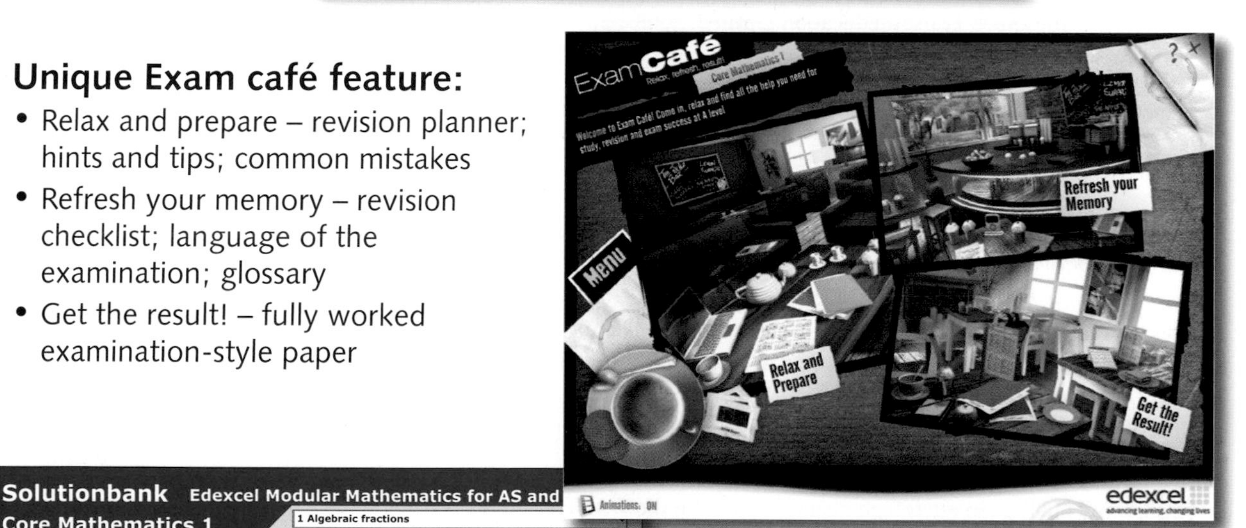

Solutionbank Edexcel Modular Mathematics for AS and A Level
Core Mathematics 1

1 Algebraic fractions
Exercises: A Questions: 1

Question:

The lines $y = -2x + 1$ and $y = x + 7$ intersect at the point L. The point M has coordinates $(-3, 1)$. Find the equation of the line that passes through the points L and M.

Hint:

Solve $y = -2x + 1$ and $y = x + 7$ simultaneously.

Solution:

$y = x + 7$

$y = -2x + 1$

So $x + 7 = -2x + 1$

Question Hint Solution Print Help

edexcel

Solutionbank

- Hints and solutions to every question in the textbook
- Solutions and commentary for all review exercises and the practice examination paper

Pearson Education Limited, a company incorporated in England and Wales, having its registered office at 80 Strand, London, WC2R 0RL. Registered company number: 872828

First published 2004 under the title Heinemann modular mathematics for Edexcel AS and A-Level Core Mathematics 1

This edition 2008

17 16
23

British Library Cataloguing in Publication Data is available from the British Library on request.

ISBN 978 0 435519 100

Edited by Susan Gardner
Typeset by Techset Ltd, Gateshead
Original illustrations © Pearson Education Limited, 2008
Illustrated by Techset Ltd, Gateshead
Cover design by Christopher Howson
Picture research by Chrissie Martin
Cover photo/illustration © Science Photo Library/Laguna Design
Printed in China (CTPS/23)

Acknowledgements
The author and publisher would like to thank the following individuals and organisations for permission to reproduce photographs:

Alamy / Hideo Kurihara p**1**; Shutterstock / photosbyjohn p**15**; Creatas p**27**; iStockPhoto.com p**41**; Pearson Education Ltd / Jules Selmes p**73**; akg-images p**91**; Science Photo Library / Sheila Terry p**112**; Bridgeman Arts Library Ltd / Private Collection p**133**

Every effort has been made to contact copyright holders of material reproduced in this book. Any omissions will be rectified in subsequent printings if notice is given to the publishers.

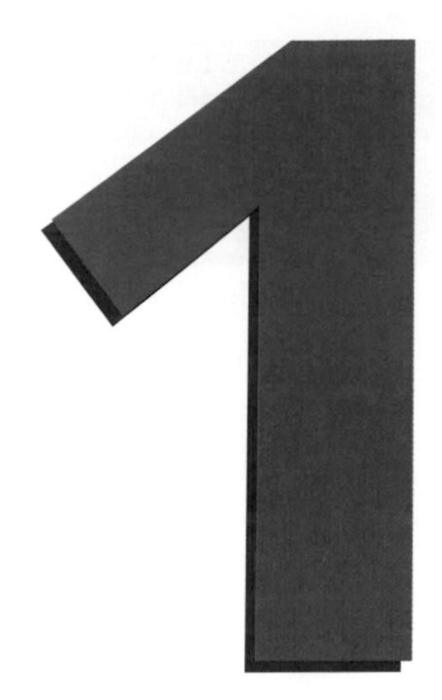

After completing this chapter you should be able to

1 simplify expressions and collect like terms

2 apply the rules of indices

3 multiply out brackets

4 factorise expressions including quadratics

5 manipulate surds.

This chapter provides the foundations for many aspects of A level Mathematics. Factorising expressions will enable you to solve equations; it could help sketch the graph of a function. A knowledge of indices is very important when differentiating and integrating. Surds are an important way of giving exact answers to problems and you will meet them again when solving quadratic equations.

Algebra and functions

Did you know?

...that the surd

$$\frac{\sqrt{5}+1}{2} \approx 1.618$$

is a number that occurs both in nature and the arts? It is called the 'golden ratio' and describes the ratio of the longest side of a rectangle to the shortest. It is supposed to be the most aesthetically pleasing rectangular shape and has been used by artists and designers since Ancient Greek times.

The Parthenon, showing the 'golden ratio' in its proportions.

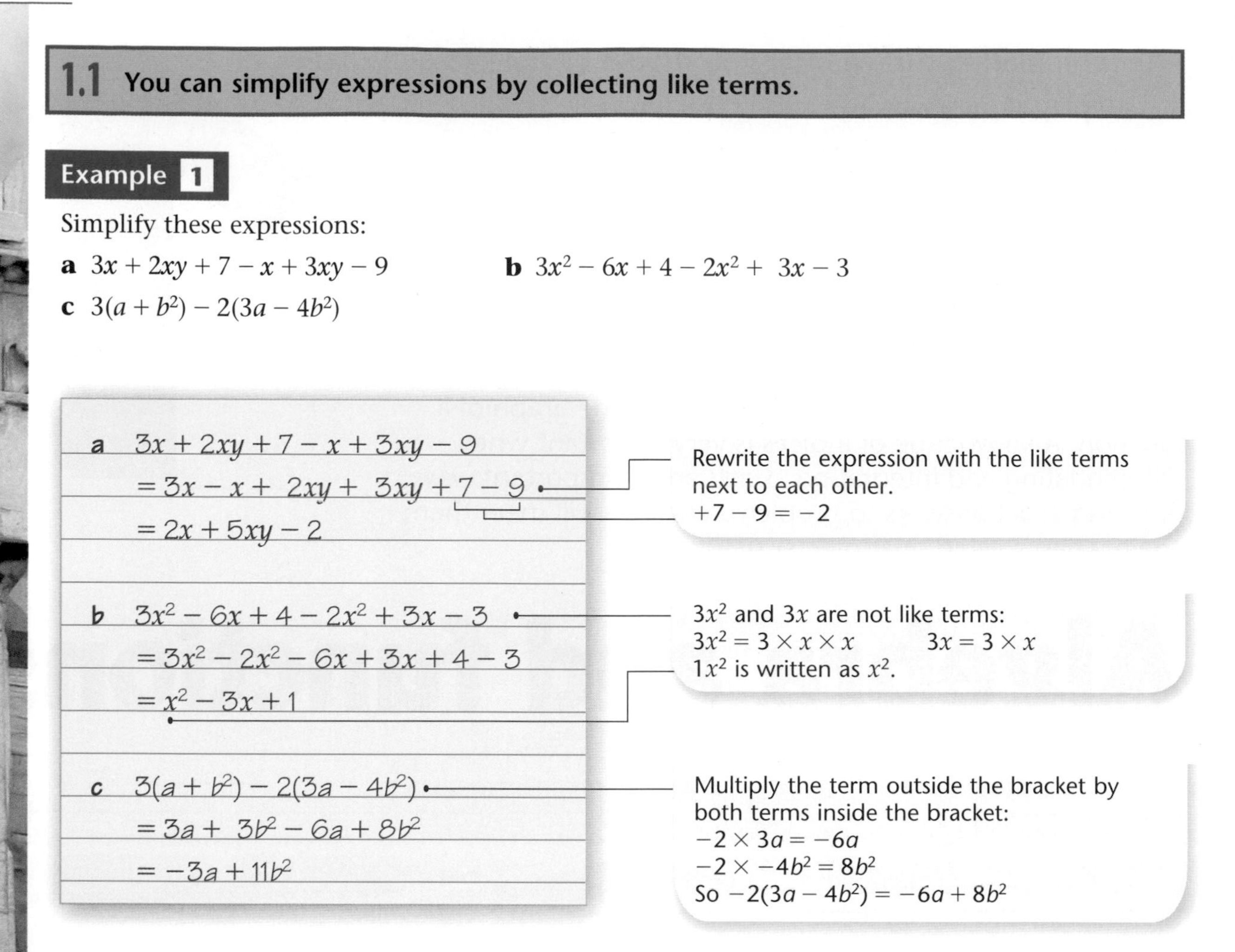

1.1 You can simplify expressions by collecting like terms.

Example 1

Simplify these expressions:

a $3x + 2xy + 7 - x + 3xy - 9$ **b** $3x^2 - 6x + 4 - 2x^2 + 3x - 3$

c $3(a + b^2) - 2(3a - 4b^2)$

a $3x + 2xy + 7 - x + 3xy - 9$

$= 3x - x + 2xy + 3xy + 7 - 9$

$= 2x + 5xy - 2$

Rewrite the expression with the like terms next to each other.

$+7 - 9 = -2$

b $3x^2 - 6x + 4 - 2x^2 + 3x - 3$

$= 3x^2 - 2x^2 - 6x + 3x + 4 - 3$

$= x^2 - 3x + 1$

$3x^2$ and $3x$ are not like terms:

$3x^2 = 3 \times x \times x$ $3x = 3 \times x$

$1x^2$ is written as x^2.

c $3(a + b^2) - 2(3a - 4b^2)$

$= 3a + 3b^2 - 6a + 8b^2$

$= -3a + 11b^2$

Multiply the term outside the bracket by both terms inside the bracket:

$-2 \times 3a = -6a$

$-2 \times -4b^2 = 8b^2$

So $-2(3a - 4b^2) = -6a + 8b^2$

Exercise 1A

Simplify these expressions:

1 $4x - 5y + 3x + 6y$

2 $3r + 7t - 5r + 3t$

3 $3m - 2n - p + 5m + 3n - 6p$

4 $3ab - 3ac + 3a - 7ab + 5ac$

5 $7x^2 - 2x^2 + 5x^2 - 4x^2$

6 $4m^2n + 5mn^2 - 2m^2n + mn^2 - 3mn^2$

7 $5x^2 + 4x + 1 - 3x^2 + 2x + 7$

8 $6x^2 + 5x - 12 + 3x^2 - 7x + 11$

9 $3x^2 - 5x + 2 + 3x^2 - 7x - 12$

10 $4c^2d + 5cd^2 - c^2d + 3cd^2 + 7c^2d$

11 $2x^2 + 3x + 1 + 2(3x^2 + 6)$

12 $4(a + a^2b) - 3(2a + a^2b)$

13 $2(3x^2 + 4x + 5) - 3(x^2 - 2x - 3)$

14 $7(1 - x^2) + 3(2 - 3x + 5x^2)$

15 $4(a + b + 3c) - 3a + 2c$

16 $4(c + 3d^2) - 3(2c + d^2)$

17 $5 - 3(x^2 + 2x - 5) + 3x^2$

18 $(r^2 + 3t^2 + 9) - (2r^2 + 3t^2 - 4)$

1.2 You can simplify expressions and functions by using rules of indices (powers).

- $a^m \times a^n = a^{m+n}$
  $a^m \div a^n = a^{m-n}$
  $(a^m)^n = a^{mn}$
  $a^{-m} = \frac{1}{a^m}$
  $a^{\frac{1}{m}} = \sqrt[m]{a}$ — The mth root of a.
  $a^{\frac{n}{m}} = \sqrt[m]{a^n}$

Example 2

Simplify these expressions:

a $x^2 \times x^5$ **b** $2r^2 \times 3r^3$ **c** $b^4 \div b^4$
d $6x^{-3} \div 3x^{-5}$ **e** $(a^3)^2 \times 2a^2$ **f** $(3x^2)^3 \div x^4$

a $x^2 \times x^5$
$= x^{2+5}$ — Use the rule $a^m \times a^n = a^{m+n}$ to simplify the index.
$= x^7$

b $2r^2 \times 3r^3$
$= 2 \times 3 \times r^2 \times r^3$ — Rewrite the expression with the numbers together and the r terms together.
$= 6 \times r^{2+3}$ — $2 \times 3 = 6$; $r^2 \times r^3 = r^{2+3}$
$= 6r^5$

c $b^4 \div b^4$ — Use the rule $a^m \div a^n = a^{m-n}$
$= b^{4-4}$
$= b^0 = 1$ — Any term raised to the power of zero = 1.

d $6x^{-3} \div 3x^{-5}$
$= 6 \div 3 \times x^{-3} \div x^{-5}$ — $x^{-3} \div x^{-5} = x^{-3--5} = x^2$
$= 2 \times x^2$
$= 2x^2$

e $(a^3)^2 \times 2a^2$ — Use the rule $(a^m)^n = a^{mn}$ to simplify the index.
$= a^6 \times 2a^2$ — $a^6 \times 2a^2 = 1 \times 2 \times a^6 \times a^2$ $= 2 \times a^{6+2}$
$= 2 \times a^6 \times a^2$
$= 2 \times a^{6+2}$
$= 2a^8$

f $(3x^2)^3 \div x^4$ — Use the rule $(a^m)^n = a^{mn}$ to simplify the index.
$= 27x^6 \div x^4$
$= 27 \div 1 \times x^6 \div x^4$
$= 27 \times x^{6-4}$
$= 27x^2$

Exercise 1B

Simplify these expressions:

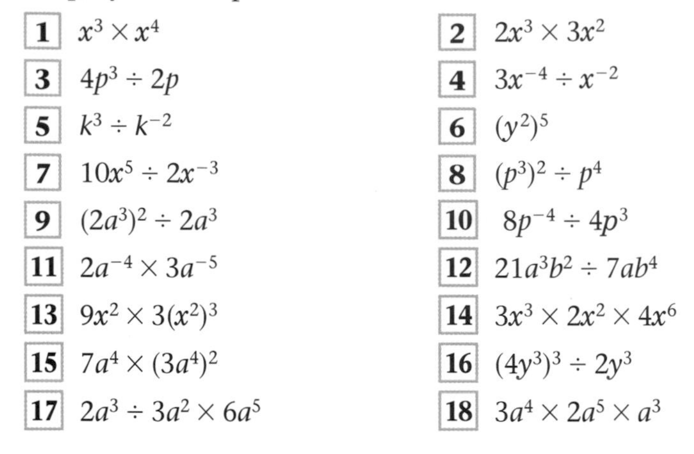

1 $x^3 \times x^4$

2 $2x^3 \times 3x^2$

3 $4p^3 \div 2p$

4 $3x^{-4} \div x^{-2}$

5 $k^3 \div k^{-2}$

6 $(y^2)^5$

7 $10x^5 \div 2x^{-3}$

8 $(p^3)^2 \div p^4$

9 $(2a^3)^2 \div 2a^3$

10 $8p^{-4} \div 4p^3$

11 $2a^{-4} \times 3a^{-5}$

12 $21a^3b^2 \div 7ab^4$

13 $9x^2 \times 3(x^2)^3$

14 $3x^3 \times 2x^2 \times 4x^6$

15 $7a^4 \times (3a^4)^2$

16 $(4y^3)^3 \div 2y^3$

17 $2a^3 \div 3a^2 \times 6a^5$

18 $3a^4 \times 2a^5 \times a^3$

1.3 You can expand an expression by multiplying each term inside the bracket by the term outside.

Example 3

Expand these expressions, simplify if possible:

a $5(2x + 3)$

b $-3x(7x - 4)$

c $y^2(3 - 2y^3)$

d $4x(3x - 2x^2 + 5x^3)$

e $2x(5x + 3) - 5(2x + 3)$

Hint: A – sign outside a bracket changes the sign of every term inside the brackets.

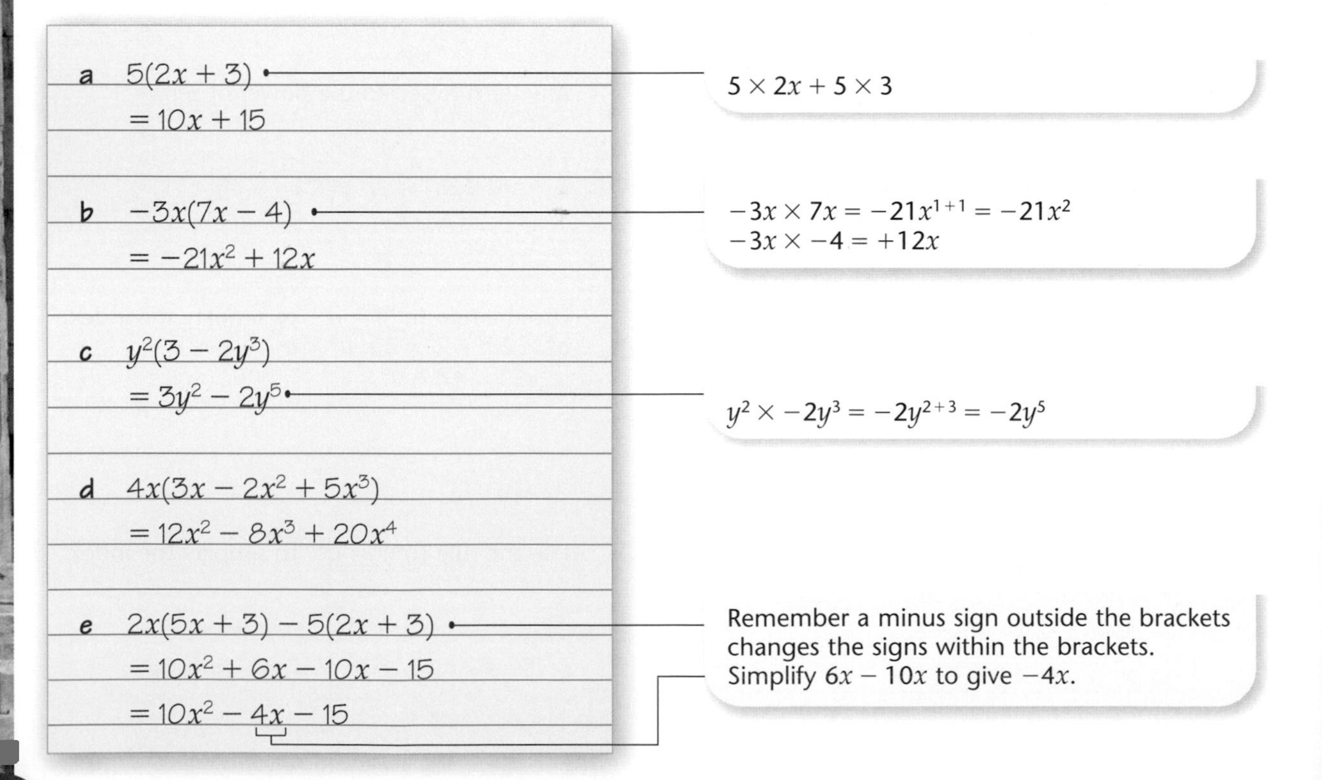

a $5(2x + 3)$
$= 10x + 15$

$5 \times 2x + 5 \times 3$

b $-3x(7x - 4)$
$= -21x^2 + 12x$

$-3x \times 7x = -21x^{1+1} = -21x^2$
$-3x \times -4 = +12x$

c $y^2(3 - 2y^3)$
$= 3y^2 - 2y^5$

$y^2 \times -2y^3 = -2y^{2+3} = -2y^5$

d $4x(3x - 2x^2 + 5x^3)$
$= 12x^2 - 8x^3 + 20x^4$

e $2x(5x + 3) - 5(2x + 3)$
$= 10x^2 + 6x - 10x - 15$
$= 10x^2 - 4x - 15$

Remember a minus sign outside the brackets changes the signs within the brackets.
Simplify $6x - 10x$ to give $-4x$.

Exercise 1C

Expand and simplify if possible:

1	$9(x - 2)$	**2**	$x(x + 9)$
3	$-3y(4 - 3y)$	**4**	$x(y + 5)$
5	$-x(3x + 5)$	**6**	$-5x(4x + 1)$
7	$(4x + 5)x$	**8**	$-3y(5 - 2y^2)$
9	$-2x(5x - 4)$	**10**	$(3x - 5)x^2$
11	$3(x + 2) + (x - 7)$	**12**	$5x - 6 - (3x - 2)$
13	$x(3x^2 - 2x + 5)$	**14**	$7y^2(2 - 5y + 3y^2)$
15	$-2y^2(5 - 7y + 3y^2)$	**16**	$7(x - 2) + 3(x + 4) - 6(x - 2)$
17	$5x - 3(4 - 2x) + 6$	**18**	$3x^2 - x(3 - 4x) + 7$
19	$4x(x + 3) - 2x(3x - 7)$	**20**	$3x^2(2x + 1) - 5x^2(3x - 4)$

1.4 You can factorise expressions.

■ **Factorising is the opposite of expanding expressions.**

When you have completely factorised an expression, the terms inside do not have a common factor.

Example 4

Factorise these expressions completely:

a $3x + 9$ **b** $x^2 - 5x$ **c** $8x^2 + 20x$

d $9x^2y + 15xy^2$ **e** $3x^2 - 9xy$

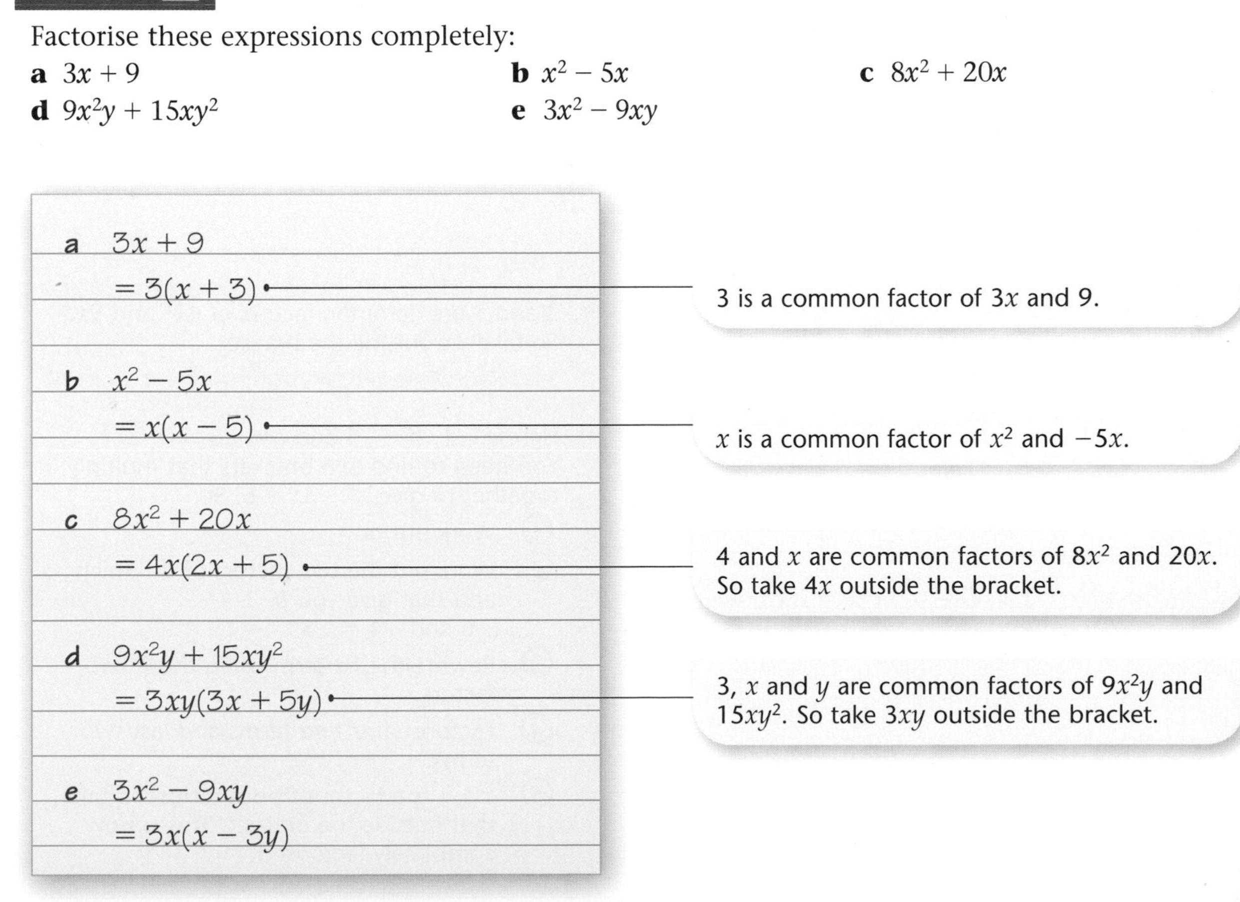

Exercise 1D

Factorise these expressions completely:

1 $4x + 8$

2 $6x - 24$

3 $20x + 15$

4 $2x^2 + 4$

5 $4x^2 + 20$

6 $6x^2 - 18x$

7 $x^2 - 7x$

8 $2x^2 + 4x$

9 $3x^2 - x$

10 $6x^2 - 2x$

11 $10y^2 - 5y$

12 $35x^2 - 28x$

13 $x^2 + 2x$

14 $3y^2 + 2y$

15 $4x^2 + 12x$

16 $5y^2 - 20y$

17 $9xy^2 + 12x^2y$

18 $6ab - 2ab^2$

19 $5x^2 - 25xy$

20 $12x^2y + 8xy^2$

21 $15y - 20yz^2$

22 $12x^2 - 30$

23 $xy^2 - x^2y$

24 $12y^2 - 4yx$

1.5 You can factorise quadratic expressions.

■ **A quadratic expression has the form $ax^2 + bx + c$, where a, b, c are constants and $a \neq 0$.**

Example 5

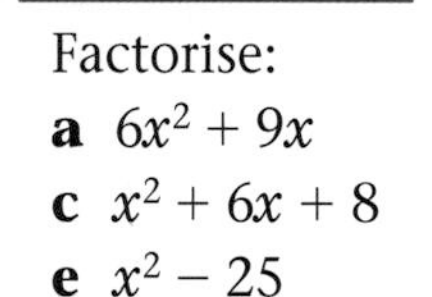

Factorise:

a $6x^2 + 9x$

b $x^2 - 5x - 6$

c $x^2 + 6x + 8$

d $6x^2 - 11x - 10$

e $x^2 - 25$

f $4x^2 - 9y^2$

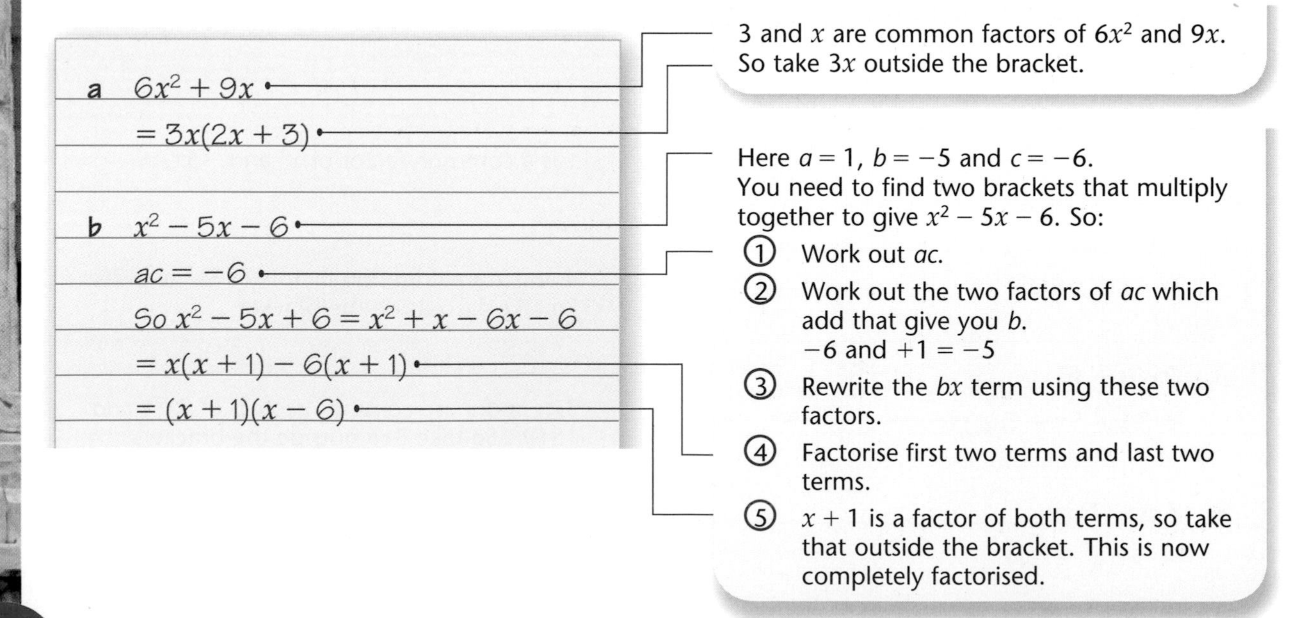

a $6x^2 + 9x$

$= 3x(2x + 3)$

3 and x are common factors of $6x^2$ and $9x$. So take $3x$ outside the bracket.

b $x^2 - 5x - 6$

$ac = -6$

So $x^2 - 5x + 6 = x^2 + x - 6x - 6$

$= x(x + 1) - 6(x + 1)$

$= (x + 1)(x - 6)$

Here $a = 1$, $b = -5$ and $c = -6$. You need to find two brackets that multiply together to give $x^2 - 5x - 6$. So:

① Work out ac.

② Work out the two factors of ac which add that give you b. -6 and $+1 = -5$

③ Rewrite the bx term using these two factors.

④ Factorise first two terms and last two terms.

⑤ $x + 1$ is a factor of both terms, so take that outside the bracket. This is now completely factorised.

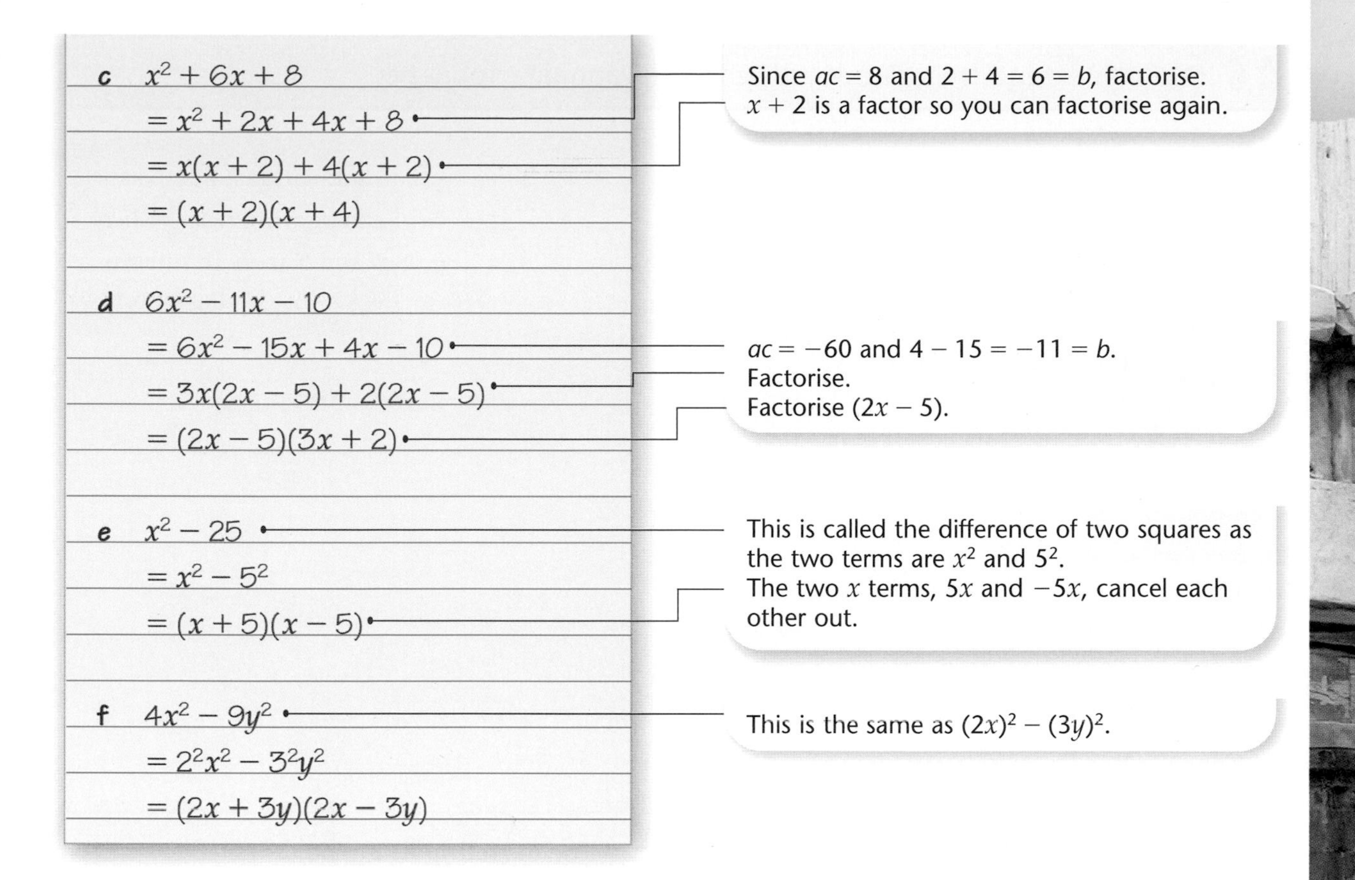

c $x^2 + 6x + 8$

$= x^2 + 2x + 4x + 8$

$= x(x + 2) + 4(x + 2)$

$= (x + 2)(x + 4)$

Since $ac = 8$ and $2 + 4 = 6 = b$, factorise.
$x + 2$ is a factor so you can factorise again.

d $6x^2 - 11x - 10$

$= 6x^2 - 15x + 4x - 10$

$= 3x(2x - 5) + 2(2x - 5)$

$= (2x - 5)(3x + 2)$

$ac = -60$ and $4 - 15 = -11 = b$.
Factorise.
Factorise $(2x - 5)$.

e $x^2 - 25$

$= x^2 - 5^2$

$= (x + 5)(x - 5)$

This is called the difference of two squares as the two terms are x^2 and 5^2.
The two x terms, $5x$ and $-5x$, cancel each other out.

f $4x^2 - 9y^2$

$= 2^2x^2 - 3^2y^2$

$= (2x + 3y)(2x - 3y)$

This is the same as $(2x)^2 - (3y)^2$.

■ $x^2 - y^2 = (x + y)(x - y)$
This is called the difference of two squares.

Exercise 1E

Factorise:

1	$x^2 + 4x$	**2**	$2x^2 + 6x$
3	$x^2 + 11x + 24$	**4**	$x^2 + 8x + 12$
5	$x^2 + 3x - 40$	**6**	$x^2 - 8x + 12$
7	$x^2 + 5x + 6$	**8**	$x^2 - 2x - 24$
9	$x^2 - 3x - 10$	**10**	$x^2 + x - 20$
11	$2x^2 + 5x + 2$	**12**	$3x^2 + 10x - 8$
13	$5x^2 - 16x + 3$	**14**	$6x^2 - 8x - 8$
15	$2x^2 + 7x - 15$	**16**	$2x^4 + 14x^2 + 24$
17	$x^2 - 4$	**18**	$x^2 - 49$
19	$4x^2 - 25$	**20**	$9x^2 - 25y^2$
21	$36x^2 - 4$	**22**	$2x^2 - 50$
23	$6x^2 - 10x + 4$	**24**	$15x^2 + 42x - 9$

Hints:
Question 14 – Take 2 out as a common factor first.
Question 16 – let $y = x^2$.

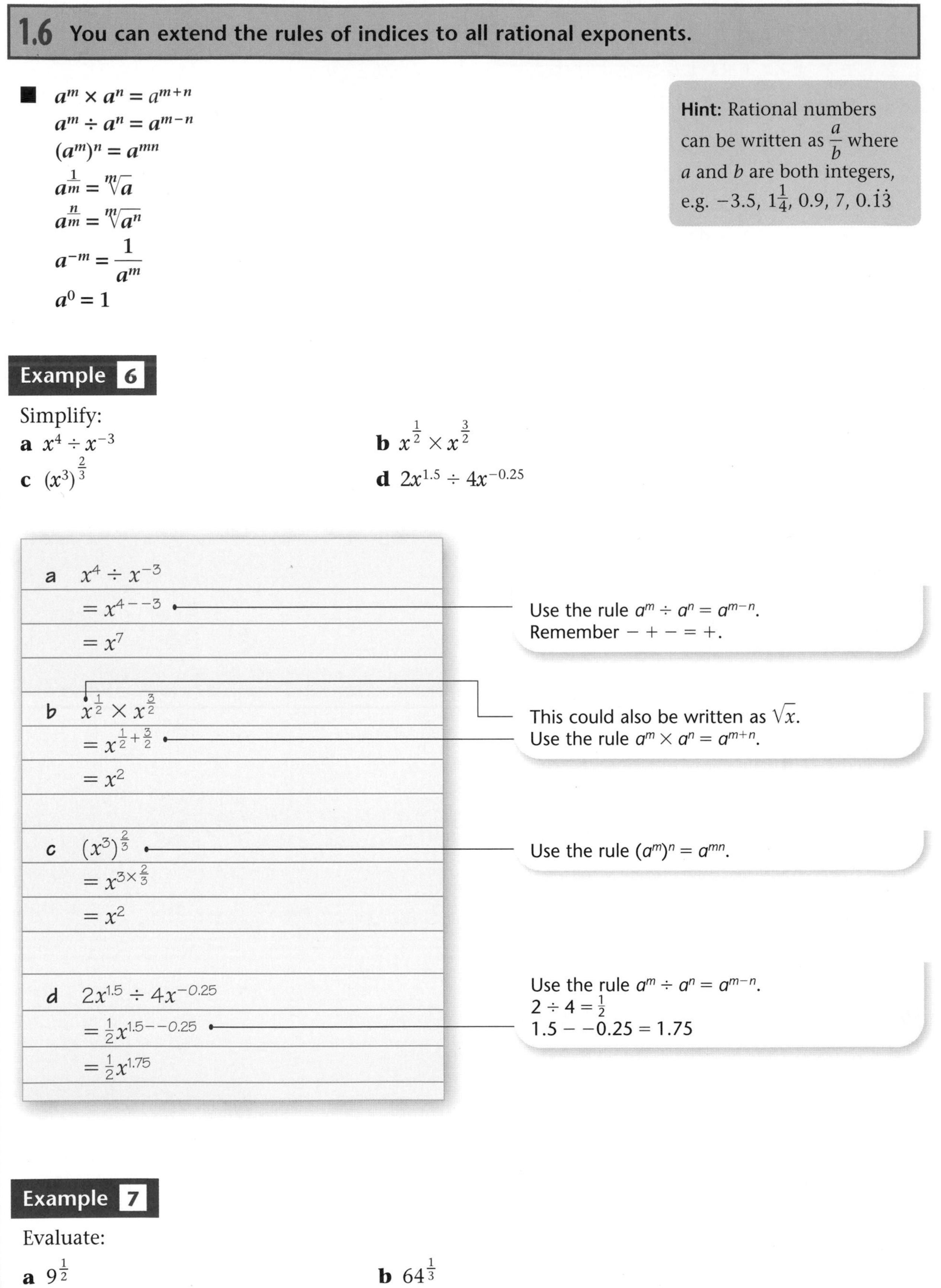

1.6 You can extend the rules of indices to all rational exponents.

■ $a^m \times a^n = a^{m+n}$

$a^m \div a^n = a^{m-n}$

$(a^m)^n = a^{mn}$

$a^{\frac{1}{m}} = \sqrt[m]{a}$

$a^{\frac{n}{m}} = \sqrt[m]{a^n}$

$a^{-m} = \dfrac{1}{a^m}$

$a^0 = 1$

Hint: Rational numbers can be written as $\frac{a}{b}$ where a and b are both integers, e.g. -3.5, $1\frac{1}{4}$, 0.9, 7, $0.\dot{1}\dot{3}$

Example 6

Simplify:

a $x^4 \div x^{-3}$

b $x^{\frac{1}{2}} \times x^{\frac{3}{2}}$

c $(x^3)^{\frac{2}{3}}$

d $2x^{1.5} \div 4x^{-0.25}$

a $x^4 \div x^{-3}$

$= x^{4--3}$

$= x^7$

Use the rule $a^m \div a^n = a^{m-n}$.
Remember $- + - = +$.

b $x^{\frac{1}{2}} \times x^{\frac{3}{2}}$

$= x^{\frac{1}{2}+\frac{3}{2}}$

$= x^2$

This could also be written as $\sqrt{x}$.
Use the rule $a^m \times a^n = a^{m+n}$.

c $(x^3)^{\frac{2}{3}}$

$= x^{3\times\frac{2}{3}}$

$= x^2$

Use the rule $(a^m)^n = a^{mn}$.

d $2x^{1.5} \div 4x^{-0.25}$

$= \frac{1}{2}x^{1.5--0.25}$

$= \frac{1}{2}x^{1.75}$

Use the rule $a^m \div a^n = a^{m-n}$.
$2 \div 4 = \frac{1}{2}$
$1.5 - -0.25 = 1.75$

Example 7

Evaluate:

a $9^{\frac{1}{2}}$

b $64^{\frac{1}{3}}$

c $49^{\frac{3}{2}}$

d $25^{-\frac{3}{2}}$

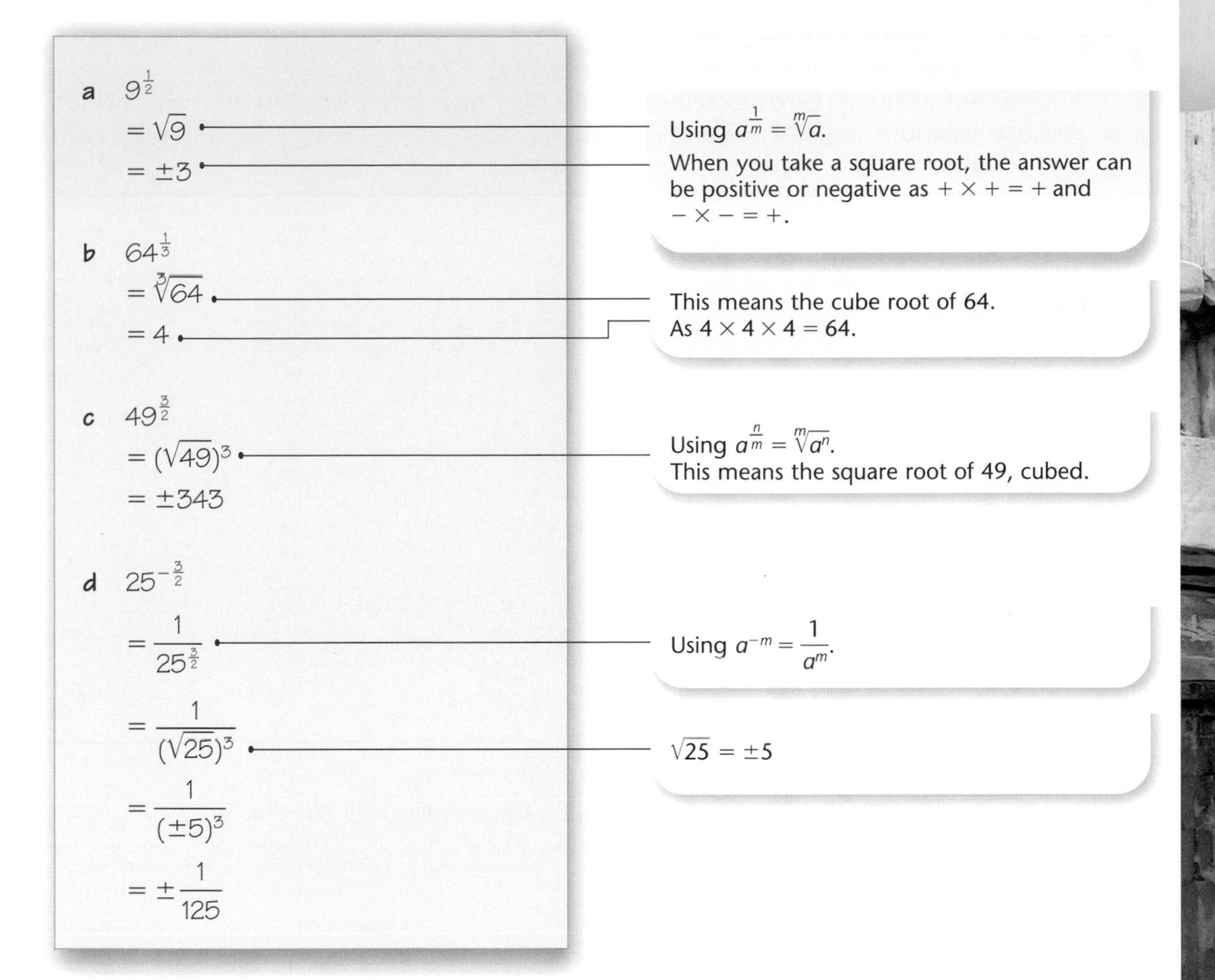

Exercise 1F

1 Simplify:

a $x^3 \div x^{-2}$ **b** $x^5 \div x^7$ **c** $x^{\frac{3}{2}} \times x^{\frac{5}{2}}$

d $(x^2)^{\frac{3}{2}}$ **e** $(x^3)^{\frac{5}{3}}$ **f** $3x^{0.5} \times 4x^{-0.5}$

g $9x^{\frac{2}{3}} \div 3x^{\frac{1}{6}}$ **h** $5x^{1\frac{2}{5}} \div x^{\frac{2}{5}}$ **i** $3x^4 \times 2x^{-5}$

2 Evaluate:

a $25^{\frac{1}{2}}$ **b** $81^{\frac{1}{2}}$ **c** $27^{\frac{1}{3}}$

d 4^{-2} **e** $9^{-\frac{1}{2}}$ **f** $(-5)^{-3}$

g $(\frac{3}{4})^0$ **h** $1296^{\frac{1}{4}}$ **i** $(1\frac{9}{16})^{\frac{3}{2}}$

j $(\frac{27}{8})^{\frac{2}{3}}$ **k** $(\frac{6}{5})^{-1}$ **l** $(\frac{343}{512})^{-\frac{2}{3}}$

1.7 **You can write a number exactly using surds, e.g. $\sqrt{2}$, $\sqrt{3} - 5$, $\sqrt{19}$.**
You cannot evaluate surds exactly because they give never-ending, non-repeating decimal fractions, e.g. $\sqrt{2} = 1.414\,213\,562...$
The square root of a prime number is a surd.

■ **You can manipulate surds using these rules:**

$\sqrt{(ab)} = \sqrt{a} \times \sqrt{b}$

$\sqrt{\frac{a}{b}} = \frac{\sqrt{a}}{\sqrt{b}}$

Example 8

Simplify:

a $\sqrt{12}$ **b** $\frac{\sqrt{20}}{2}$ **c** $5\sqrt{6} - 2\sqrt{24} + \sqrt{294}$

a $\sqrt{12}$

$= \sqrt{(4 \times 3)}$

$= \sqrt{4} \times \sqrt{3}$ — Use the rule $\sqrt{ab} = \sqrt{a} \times \sqrt{b}$. $\sqrt{4} = 2$

$= 2\sqrt{3}$

b $\frac{\sqrt{20}}{2}$ — $\sqrt{20} = \sqrt{4} \times \sqrt{5}$

$= \frac{\sqrt{4} \times \sqrt{5}}{2}$ — $\sqrt{4} = 2$

$= \frac{2 \times \sqrt{5}}{2}$ — Cancel by 2.

$= \sqrt{5}$

c $5\sqrt{6} - 2\sqrt{24} + \sqrt{294}$

$= 5\sqrt{6} - 2\sqrt{6}\sqrt{4} + \sqrt{6} \times \sqrt{49}$ — $\sqrt{6}$ is a common factor.

$= \sqrt{6}(5 - 2\sqrt{4} + \sqrt{49})$ — Work out the square roots $\sqrt{4}$ and $\sqrt{49}$.

$= \sqrt{6}(5 - 2 \times 2 + 7)$ — $5 - 4 + 7 = 8$

$= \sqrt{6}(8)$

$= 8\sqrt{6}$

Exercise 1G

Simplify:

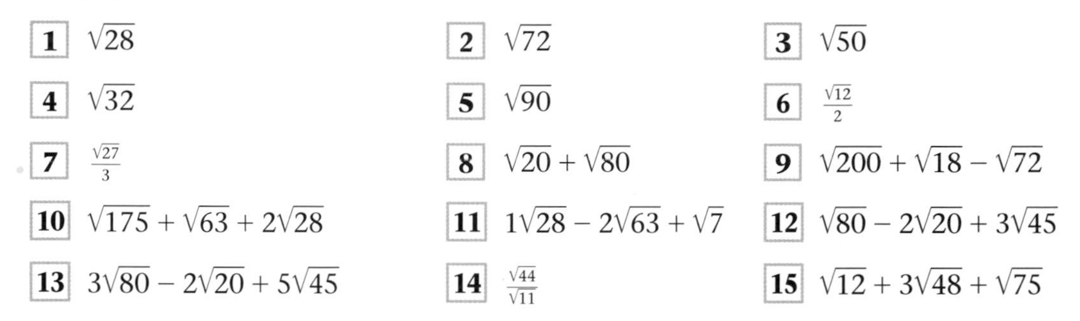

1 $\sqrt{28}$ **2** $\sqrt{72}$ **3** $\sqrt{50}$

4 $\sqrt{32}$ **5** $\sqrt{90}$ **6** $\frac{\sqrt{12}}{2}$

7 $\frac{\sqrt{27}}{3}$ **8** $\sqrt{20} + \sqrt{80}$ **9** $\sqrt{200} + \sqrt{18} - \sqrt{72}$

10 $\sqrt{175} + \sqrt{63} + 2\sqrt{28}$ **11** $1\sqrt{28} - 2\sqrt{63} + \sqrt{7}$ **12** $\sqrt{80} - 2\sqrt{20} + 3\sqrt{45}$

13 $3\sqrt{80} - 2\sqrt{20} + 5\sqrt{45}$ **14** $\frac{\sqrt{44}}{\sqrt{11}}$ **15** $\sqrt{12} + 3\sqrt{48} + \sqrt{75}$

1.8 You rationalise the denominator of a fraction when it is a surd.

■ **The rules to rationalise surds are:**

- **Fractions in the form $\frac{1}{\sqrt{a}}$, multiply the top and bottom by $\sqrt{a}$.**
- **Fractions in the form $\frac{1}{a + \sqrt{b}}$, multiply the top and bottom by $a - \sqrt{b}$.**
- **Fractions in the form $\frac{1}{a - \sqrt{b}}$, multiply the top and bottom by $a + \sqrt{b}$.**

Example 9

Rationalise the denominator of:

a $\frac{1}{\sqrt{3}}$ **b** $\frac{1}{3 + \sqrt{2}}$ **c** $\frac{\sqrt{5} + \sqrt{2}}{\sqrt{5} - \sqrt{2}}$

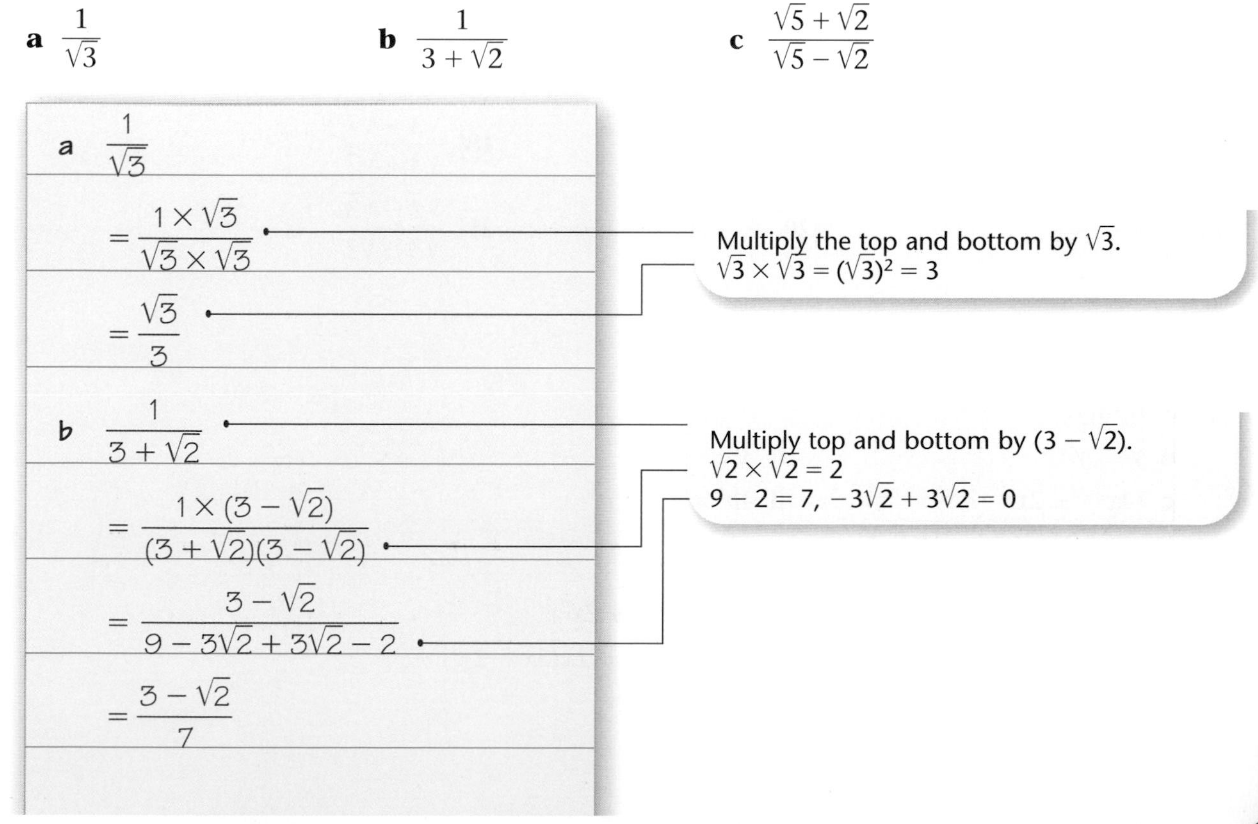

a $\frac{1}{\sqrt{3}}$

$= \frac{1 \times \sqrt{3}}{\sqrt{3} \times \sqrt{3}}$

$= \frac{\sqrt{3}}{3}$

Multiply the top and bottom by $\sqrt{3}$.
$\sqrt{3} \times \sqrt{3} = (\sqrt{3})^2 = 3$

b $\frac{1}{3 + \sqrt{2}}$

$= \frac{1 \times (3 - \sqrt{2})}{(3 + \sqrt{2})(3 - \sqrt{2})}$

$= \frac{3 - \sqrt{2}}{9 - 3\sqrt{2} + 3\sqrt{2} - 2}$

$= \frac{3 - \sqrt{2}}{7}$

Multiply top and bottom by $(3 - \sqrt{2})$.
$\sqrt{2} \times \sqrt{2} = 2$
$9 - 2 = 7$, $-3\sqrt{2} + 3\sqrt{2} = 0$

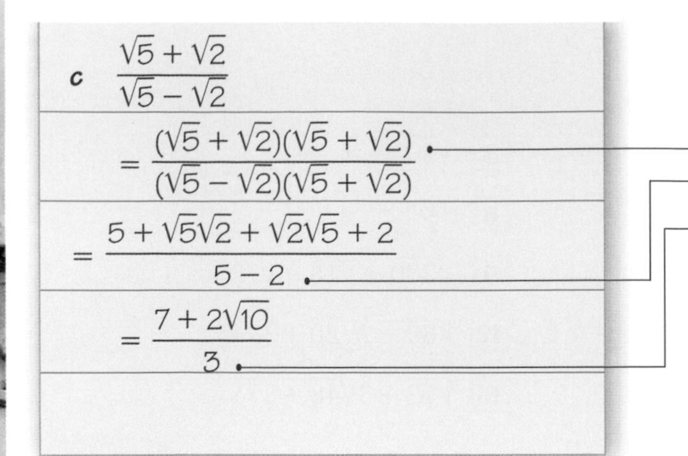

Multiply top and bottom by $\sqrt{5}+\sqrt{2}$.
$-\sqrt{2}\sqrt{5}$ and $\sqrt{5}\sqrt{2}$ cancel each other out.
$\sqrt{5}\sqrt{2} = \sqrt{10}$

Exercise 1H

Rationalise the denominators and simplify:

1 $\frac{1}{\sqrt{5}}$ **2** $\frac{1}{\sqrt{11}}$ **3** $\frac{1}{\sqrt{2}}$

4 $\frac{\sqrt{3}}{\sqrt{15}}$ **5** $\frac{\sqrt{12}}{\sqrt{48}}$ **6** $\frac{\sqrt{5}}{\sqrt{80}}$

7 $\frac{\sqrt{12}}{\sqrt{156}}$ **8** $\frac{\sqrt{7}}{\sqrt{63}}$ **9** $\frac{1}{1+\sqrt{3}}$

10 $\frac{1}{2+\sqrt{5}}$ **11** $\frac{1}{3-\sqrt{7}}$ **12** $\frac{4}{3-\sqrt{5}}$

13 $\frac{1}{\sqrt{5}-\sqrt{3}}$ **14** $\frac{3-\sqrt{2}}{4-\sqrt{5}}$ **15** $\frac{5}{2+\sqrt{5}}$

16 $\frac{5\sqrt{2}}{\sqrt{8}-\sqrt{7}}$ **17** $\frac{11}{3+\sqrt{11}}$ **18** $\frac{\sqrt{3}-\sqrt{7}}{\sqrt{3}+\sqrt{7}}$

19 $\frac{\sqrt{17}-\sqrt{11}}{\sqrt{17}+\sqrt{11}}$ **20** $\frac{\sqrt{41}+\sqrt{29}}{\sqrt{41}-\sqrt{29}}$ **21** $\frac{\sqrt{2}-\sqrt{3}}{\sqrt{3}-\sqrt{2}}$

Mixed exercise 1I

1 Simplify:

a $y^3 \times y^5$ **b** $3x^2 \times 2x^5$

c $(4x^2)^3 \div 2x^5$ **d** $4b^2 \times 3b^3 \times b^4$

2 Expand the brackets:

a $3(5y + 4)$ **b** $5x^2(3 - 5x + 2x^2)$

c $5x(2x + 3) - 2x(1 - 3x)$ **d** $3x^2(1 + 3x) - 2x(3x - 2)$

3 Factorise these expressions completely:

a $3x^2 + 4x$ **b** $4y^2 + 10y$

c $x^2 + xy + xy^2$ **d** $8xy^2 + 10x^2y$

4 Factorise:

a $x^2 + 3x + 2$

b $3x^2 + 6x$

c $x^2 - 2x - 35$

d $2x^2 - x - 3$

e $5x^2 - 13x - 6$

f $6 - 5x - x^2$

5 Simplify:

a $9x^3 \div 3x^{-3}$

b $(4^{\frac{3}{2}})^{\frac{1}{3}}$

c $3x^{-2} \times 2x^4$

d $3x^{\frac{1}{3}} \div 6x^{\frac{2}{3}}$

6 Evaluate:

a $\left(\frac{8}{27}\right)^{\frac{2}{3}}$

b $\left(\frac{225}{289}\right)^{\frac{3}{2}}$

7 Simplify:

a $\frac{3}{\sqrt{63}}$

b $\sqrt{20} + 2\sqrt{45} - \sqrt{80}$

8 Rationalise:

a $\frac{1}{\sqrt{3}}$

b $\frac{1}{\sqrt{2} - 1}$

c $\frac{3}{\sqrt{3} - 2}$

d $\frac{\sqrt{23} - \sqrt{37}}{\sqrt{23} + \sqrt{37}}$

Summary of key points

1 You can simplify expressions by collecting like terms.

2 You can simplify expressions by using rules of indices (powers).

$$a^m \times a^n = a^{m+n}$$

$$a^m \div a^n = a^{m-n}$$

$$a^{-m} = \frac{1}{a^m}$$

$$a^{\frac{1}{m}} = \sqrt[m]{a}$$

$$a^{\frac{n}{m}} = \sqrt[m]{a^n}$$

$$(a^m)^n = a^{mn}$$

$$a^0 = 1$$

3 You can expand an expression by multiplying each term inside the bracket by the term outside.

4 Factorising expressions is the opposite of expanding expressions.

5 A quadratic expression has the form $ax^2 + bx + c$, where a, b, c are constants and $a \neq 0$.

6 $x^2 - y^2 = (x + y)(x - y)$
This is called a difference of squares.

7 You can write a number exactly using surds.

8 The square root of a prime number is a surd.

9 You can manipulate surds using the rules:

$$\sqrt{ab} = \sqrt{a} \times \sqrt{b}$$

$$\sqrt{\frac{a}{b}} = \frac{\sqrt{a}}{\sqrt{b}}$$

10 The rules to rationalise surds are:

- Fractions in the form $\frac{1}{\sqrt{a}}$, multiply the top and bottom by $\sqrt{a}$.
- Fractions in the form $\frac{1}{a + \sqrt{b}}$, multiply the top and bottom by $a - \sqrt{b}$.
- Fractions in the form $\frac{1}{a - \sqrt{b}}$, multiply the top and bottom by $a + \sqrt{b}$.

After completing this chapter you should be able to

1 plot the graph of a quadratic function
2 solve a quadratic function using factorisation
3 complete the square of a quadratic function
4 solve a quadratic equation by using the quadratic formula
5 calculate the discriminant of a quadratic expression
6 sketch the graph of a quadratic function.

The above techniques will enable you to solve many types of equation and inequality. The ability to spot and solve a quadratic equation is extremely important in A level Mathematics.

Quadratic functions

Did you know?

...that the path of a golf ball can be modelled by a quadratic function?

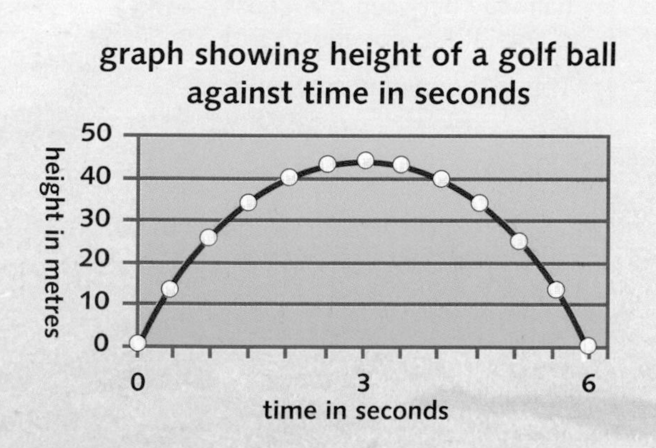

2.1 You need to be able to plot graphs of quadratic equations.

- **The general form of a quadratic equation is**

  $y = ax^2 + bx + c$

  where a, b and c are constants and $a \neq 0$.

This could also be written as $f(x) = ax^2 + bx + c$.

Example 1

a Draw the graph with equation $y = x^2 - 3x - 4$ for values of x from -2 to $+5$.

b Write down the minimum value of y and the value of x for this point.

c Label the line of symmetry.

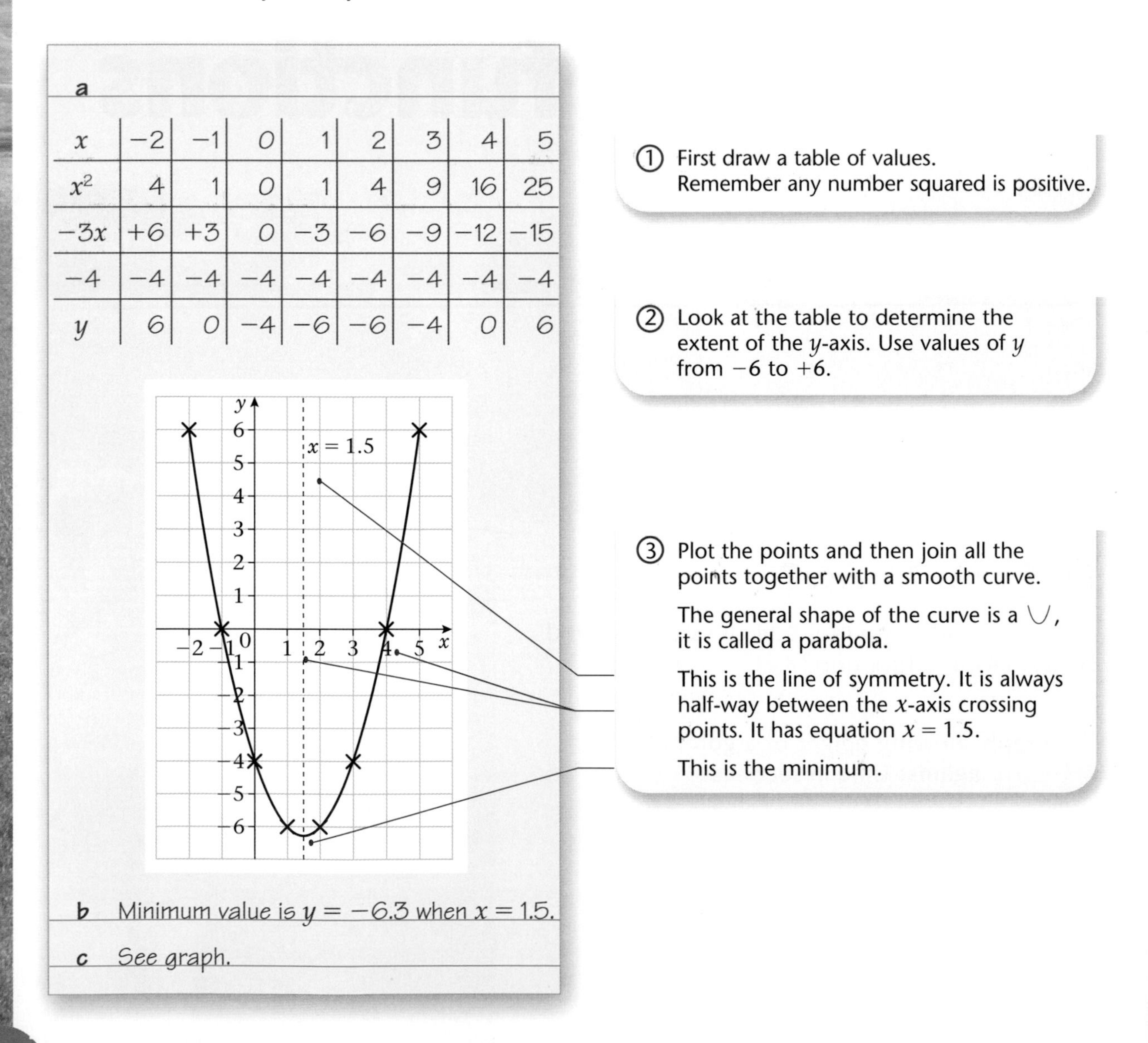

a

x	-2	-1	0	1	2	3	4	5
x^2	4	1	0	1	4	9	16	25
$-3x$	$+6$	$+3$	0	-3	-6	-9	-12	-15
-4	-4	-4	-4	-4	-4	-4	-4	-4
y	6	0	-4	-6	-6	-4	0	6

b Minimum value is $y = -6.3$ when $x = 1.5$.

c See graph.

Exercise 2A

Draw graphs with the following equations, taking values of x from -4 to $+4$.

For each graph write down the equation of the line of symmetry.

1 $y = x^2 - 3$

2 $y = x^2 + 5$

3 $y = \frac{1}{2}x^2$

4 $y = -x^2$

5 $y = (x - 1)^2$

6 $y = x^2 + 3x + 2$

7 $y = 2x^2 + 3x - 5$

8 $y = x^2 + 2x - 6$

9 $y = (2x + 1)^2$

Hint: The general shape for question **4** is an upside down ∪-shape. i.e. ∩.

2.2 You can solve quadratic equations using factorisation.

Quadratic equations have two solutions or roots. (In some cases the two roots are equal.)
To solve a quadratic equation, put it in the form $ax^2 + bx + c = 0$.

Example 2

Solve the equation $x^2 = 9x$

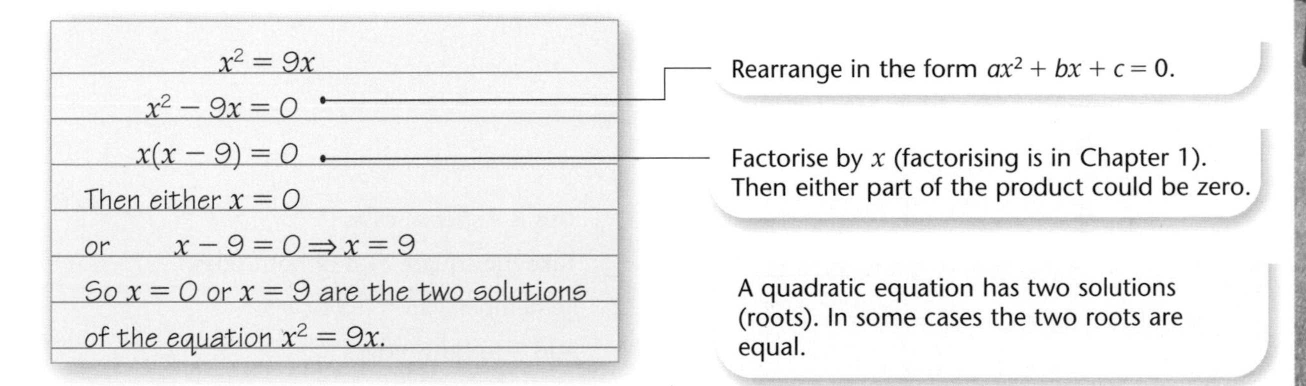

Example 3

Solve the equation $x^2 - 2x - 15 = 0$

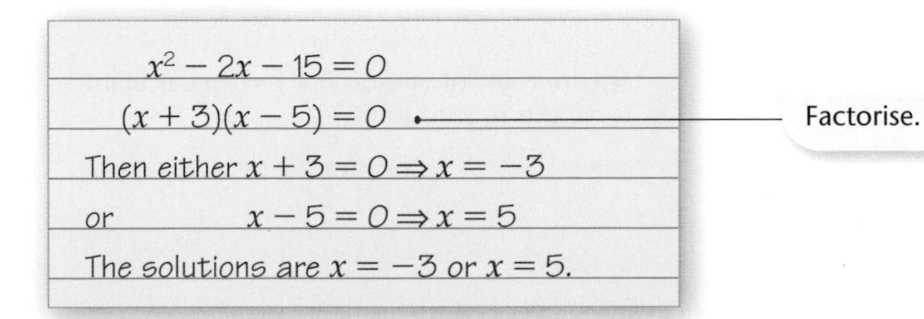

Factorise.

Example 4

Solve the equation $6x^2 + 13x - 5 = 0$

$$6x^2 + 13x - 5 = 0$$

$$(3x - 1)(2x + 5) = 0$$

Factorise.

Then either $3x - 1 = 0 \Rightarrow x = \frac{1}{3}$

or $2x + 5 = 0 \Rightarrow x = -\frac{5}{2}$

The solutions are $x = \frac{1}{3}$ or $x = -\frac{5}{2}$.

The solutions can be fractions or any other type of number.

Example 5

Solve the equation $x^2 - 5x + 18 = 2 + 3x$

$$x^2 - 5x + 18 = 2 + 3x$$

$$x^2 - 8x + 16 = 0$$

Rearrange in the form $ax^2 + bx + c = 0$.

$$(x - 4)(x - 4) = 0$$

Factorise.

Then either $x - 4 = 0 \Rightarrow x = 4$

or $x - 4 = 0 \Rightarrow x = 4$

$\Rightarrow$ $x = 4$

Here $x = 4$ is the only solution, i.e. the two roots are equal.

Example 6

Solve the equation $(2x - 3)^2 = 25$

$$(2x - 3)^2 = 25$$

$$2x - 3 = \pm 5$$

$$2x = 3 \pm 5$$

This is a special case.

Take the square root of both sides.

Remember $\sqrt{25} = +5$ or -5.

Add 3 to both sides.

Then either $2x = 3 + 5 \Rightarrow x = 4$

or $2x = 3 - 5 \Rightarrow x = -1$

The solutions are $x = 4$ or $x = -1$.

Example 7

Solve the equation $(x - 3)^2 = 7$

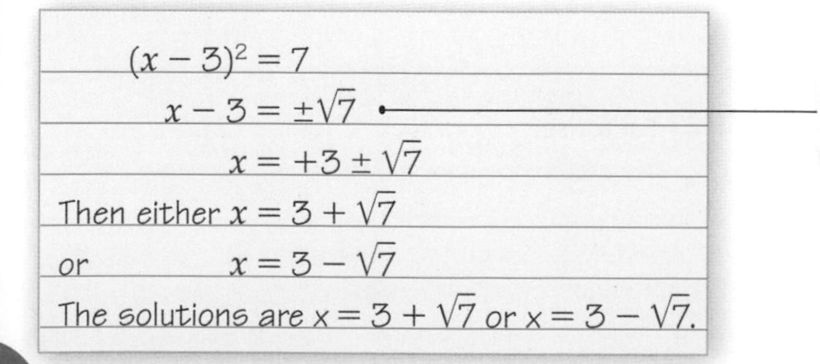

$$(x - 3)^2 = 7$$

$$x - 3 = \pm\sqrt{7}$$

$$x = +3 \pm \sqrt{7}$$

Then either $x = 3 + \sqrt{7}$

or $x = 3 - \sqrt{7}$

The solutions are $x = 3 + \sqrt{7}$ or $x = 3 - \sqrt{7}$.

Square root. (If you do not have a calculator, leave this in surd form.)

Exercise 2B

Solve the following equations:

1 $x^2 = 4x$

2 $x^2 = 25x$

3 $3x^2 = 6x$

4 $5x^2 = 30x$

5 $x^2 + 3x + 2 = 0$

6 $x^2 + 5x + 4 = 0$

7 $x^2 + 7x + 10 = 0$

8 $x^2 - x - 6 = 0$

9 $x^2 - 8x + 15 = 0$

10 $x^2 - 9x + 20 = 0$

11 $x^2 - 5x - 6 = 0$

12 $x^2 - 4x - 12 = 0$

13 $2x^2 + 7x + 3 = 0$

14 $6x^2 - 7x - 3 = 0$

15 $6x^2 - 5x - 6 = 0$

16 $4x^2 - 16x + 15 = 0$

17 $3x^2 + 5x = 2$

18 $(2x - 3)^2 = 9$

19 $(x - 7)^2 = 36$

20 $2x^2 = 8$

21 $3x^2 = 5$

22 $(x - 3)^2 = 13$

23 $(3x - 1)^2 = 11$

24 $5x^2 - 10x^2 = -7 + x + x^2$

25 $6x^2 - 7 = 11x$

26 $4x^2 + 17x = 6x - 2x^2$

2.3 You can write quadratic expressions in another form by completing the square.

$$x^2 + 2bx + b^2 = (x + b)^2$$
$$x^2 - 2bx + b^2 = (x - b)^2$$

These are both perfect squares.

To complete the square of the function $x^2 + 2bx$ you need a further term b^2. So the completed square form is

$$x^2 + 2bx = (x + b)^2 - b^2$$

Similarly

$$x^2 - 2bx = (x - b)^2 - b^2$$

Example 8

Complete the square for the expression $x^2 + 8x$

$2b = 8$, so $b = 4$

In general

■ **Completing the square:** $x^2 + bx = \left(x + \frac{b}{2}\right)^2 - \left(\frac{b}{2}\right)^2$

Example 9

Complete the square for the expressions

a $x^2 + 12x$ **b** $2x^2 - 10x$

a $x^2 + 12x$
$= (x + 6)^2 - 6^2$
$= (x + 6)^2 - 36$

$2b = 12$, so $b = 6$

b $2x^2 - 10x$
$= 2(x^2 - 5x)$
$= 2[(x - \frac{5}{2})^2 - (\frac{5}{2})^2]$
$= 2(x - \frac{5}{2})^2 - \frac{25}{2}$

Here the coefficient of x^2 is 2.
So take out the coefficient of x^2.
Complete the square on $(x^2 - 5x)$.
Use $b = -5$.

Exercise 2C

Complete the square for the expressions:

1 $x^2 + 4x$	**2** $x^2 - 6x$	**3** $x^2 - 16x$	**4** $x^2 + x$
5 $x^2 - 14x$	**6** $2x^2 + 16x$	**7** $3x^2 - 24x$	**8** $2x^2 - 4x$
9 $5x^2 + 20x$	**10** $2x^2 - 5x$	**11** $3x^2 + 9x$	**12** $3x^2 - x$

2.4 You can solve quadratic equations by completing the square.

Example 10

Solve the equation $x^2 + 8x + 10 = 0$ by completing the square.

$x^2 + 8x + 10 = 0$ — Check coefficient of $x^2 = 1$.

$x^2 + 8x = -10$ — Subtract 10 to get LHS in the form $ax^2 + b$.

$(x + 4)^2 - 4^2 = -10$ — Complete the square for $(x^2 + 8x)$.

$(x + 4)^2 = -10 + 16$ — Add 4^2 to both sides.

$(x + 4)^2 = 6$

$(x + 4) = \pm\sqrt{6}$ — Square root both sides.

$x = -4 \pm \sqrt{6}$ — Subtract 4 from both sides.

Then the solutions (roots) of
$x^2 + 8x + 10 = 0$ are either
$x = -4 + \sqrt{6}$ or $x = -4 - \sqrt{6}$.

Leave your answer in surd form as this is a non-calculator question.

Example 11

Solve the equation $2x^2 - 8x + 7 = 0$.

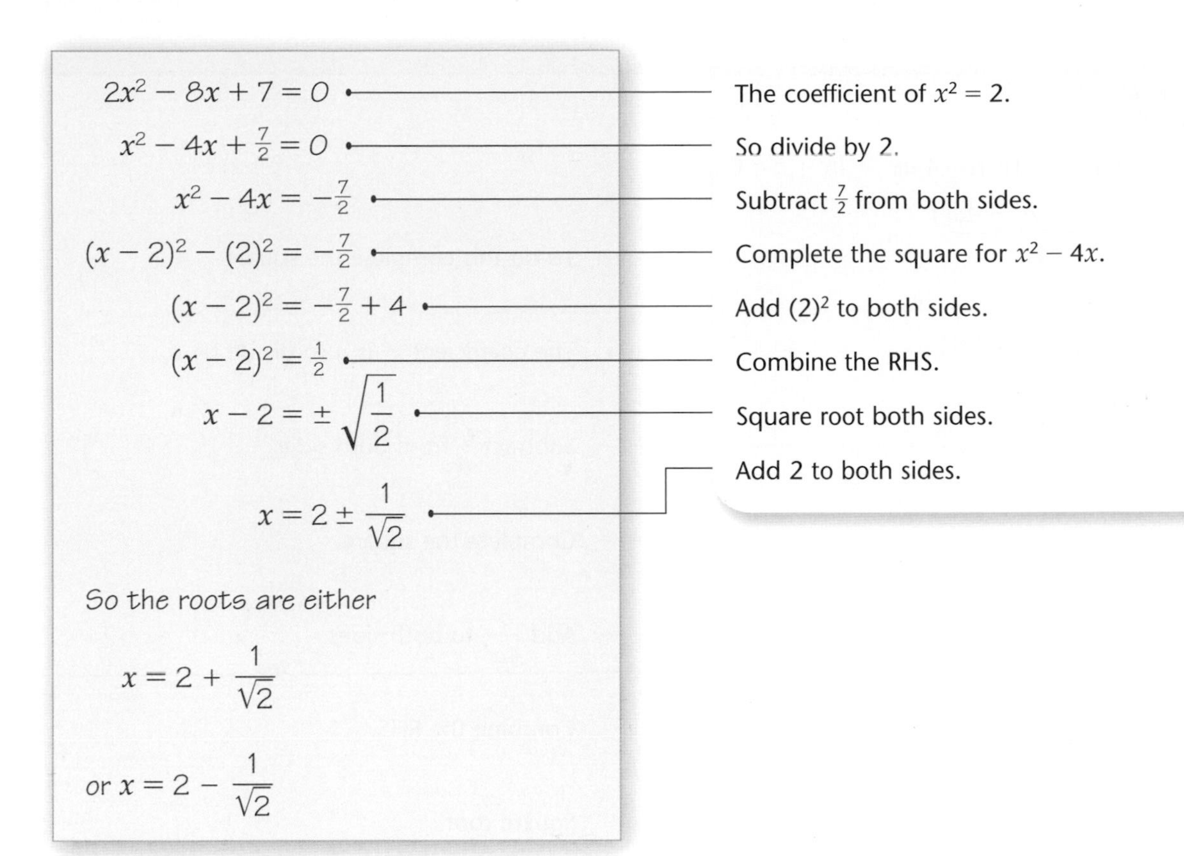
$2x^2 - 8x + 7 = 0$ — The coefficient of $x^2 = 2$.

$x^2 - 4x + \frac{7}{2} = 0$ — So divide by 2.

$x^2 - 4x = -\frac{7}{2}$ — Subtract $\frac{7}{2}$ from both sides.

$(x - 2)^2 - (2)^2 = -\frac{7}{2}$ — Complete the square for $x^2 - 4x$.

$(x - 2)^2 = -\frac{7}{2} + 4$ — Add $(2)^2$ to both sides.

$(x - 2)^2 = \frac{1}{2}$ — Combine the RHS.

$x - 2 = \pm\sqrt{\frac{1}{2}}$ — Square root both sides.

$x = 2 \pm \frac{1}{\sqrt{2}}$ — Add 2 to both sides.

So the roots are either

$x = 2 + \frac{1}{\sqrt{2}}$

or $x = 2 - \frac{1}{\sqrt{2}}$

Note: Sometimes $b^2 - 4ac$ is negative, and there are then no real solutions.

Exercise 2D

Solve these quadratic equations by completing the square (remember to leave your answer in surd form):

1 $x^2 + 6x + 1 = 0$

2 $x^2 + 12x + 3 = 0$

3 $x^2 - 10x = 5$

4 $x^2 + 4x - 2 = 0$

5 $x^2 - 3x - 5 = 0$

6 $2x^2 - 7 = 4x$

7 $4x^2 - x = 8$

8 $10 = 3x - x^2$

9 $15 - 6x - 2x^2 = 0$

10 $5x^2 + 8x - 2 = 0$

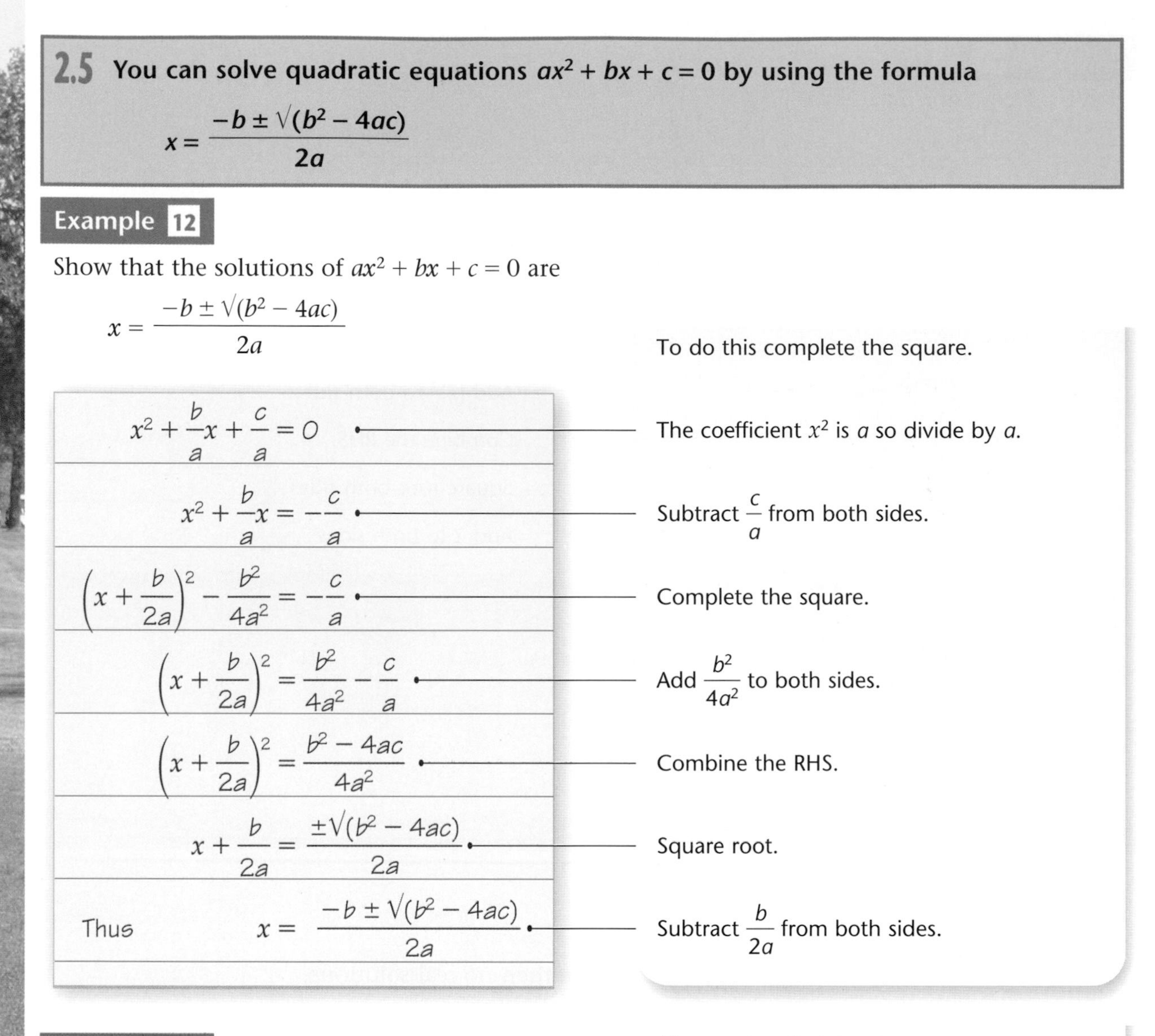

2.5 You can solve quadratic equations $ax^2 + bx + c = 0$ by using the formula

$$x = \frac{-b \pm \sqrt{(b^2 - 4ac)}}{2a}$$

Example 12

Show that the solutions of $ax^2 + bx + c = 0$ are

$$x = \frac{-b \pm \sqrt{(b^2 - 4ac)}}{2a}$$

To do this complete the square.

$$x^2 + \frac{b}{a}x + \frac{c}{a} = 0$$ — The coefficient x^2 is a so divide by a.

$$x^2 + \frac{b}{a}x = -\frac{c}{a}$$ — Subtract $\frac{c}{a}$ from both sides.

$$\left(x + \frac{b}{2a}\right)^2 - \frac{b^2}{4a^2} = -\frac{c}{a}$$ — Complete the square.

$$\left(x + \frac{b}{2a}\right)^2 = \frac{b^2}{4a^2} - \frac{c}{a}$$ — Add $\frac{b^2}{4a^2}$ to both sides.

$$\left(x + \frac{b}{2a}\right)^2 = \frac{b^2 - 4ac}{4a^2}$$ — Combine the RHS.

$$x + \frac{b}{2a} = \frac{\pm\sqrt{(b^2 - 4ac)}}{2a}$$ — Square root.

Thus $$x = \frac{-b \pm \sqrt{(b^2 - 4ac)}}{2a}$$ — Subtract $\frac{b}{2a}$ from both sides.

$b^2 - 4ac$ is called the discriminant.

Example 13

Solve $4x^2 - 3x - 2 = 0$ by using the formula.

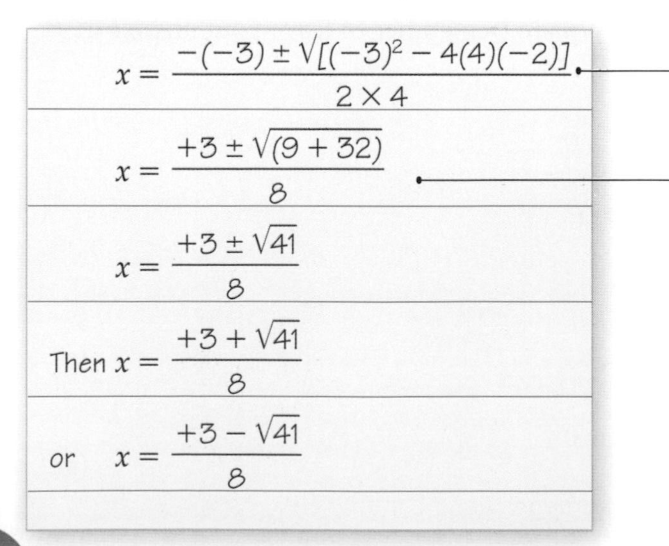

$$x = \frac{-(-3) \pm \sqrt{[(-3)^2 - 4(4)(-2)]}}{2 \times 4}$$ — Use $x = \frac{-b \pm \sqrt{(b^2 - 4ac)}}{2a}$ where $a = 4$, $b = -3$, $c = -2$.

$$x = \frac{+3 \pm \sqrt{(9 + 32)}}{8}$$ — $-4 \times 4 \times -2 = +32$

$$x = \frac{+3 \pm \sqrt{41}}{8}$$

Then $$x = \frac{+3 + \sqrt{41}}{8}$$

or $$x = \frac{+3 - \sqrt{41}}{8}$$ — Leave your answer in surd form.

Exercise 2E

Solve the following quadratic equations by using the formula, giving the solutions in surd form. Simplify your answers.

1	$x^2 + 3x + 1 = 0$	**2**	$x^2 - 3x - 2 = 0$
3	$x^2 + 6x + 6 = 0$	**4**	$x^2 - 5x - 2 = 0$
5	$3x^2 + 10x - 2 = 0$	**6**	$4x^2 - 4x - 1 = 0$
7	$7x^2 + 9x + 1 = 0$	**8**	$5x^2 + 4x - 3 = 0$
9	$4x^2 - 7x = 2$	**10**	$11x^2 + 2x - 7 = 0$

2.6 You need to be able to sketch graphs of quadratic equations and solve problems using the discriminant.

The steps to help you sketch the graphs are:

1 Decide on the shape.
When a is >0 the curve will be a ∪ shape.
When a is <0 the curve will be a ∩ shape.

2 Work out the points where the curve crosses the x- and y-axes.
Put $y = 0$ to find the x-axis crossing points coordinates.
Put $x = 0$ to find the y-axis crossing point coordinates.

3 Check the general shape of curve by considering the discriminant, $b^2 - 4ac$.
When specific conditions apply, the general shape of the curve takes these forms:

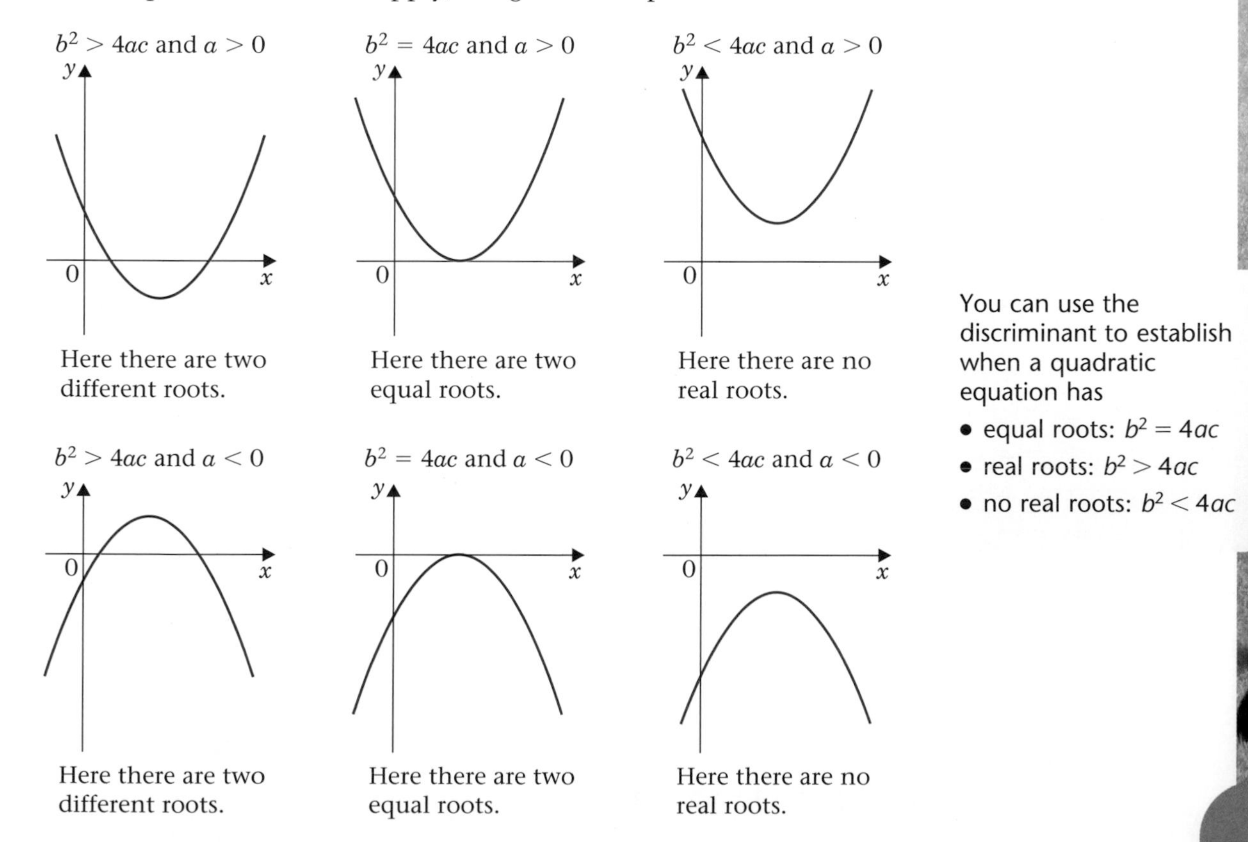

You can use the discriminant to establish when a quadratic equation has

- equal roots: $b^2 = 4ac$
- real roots: $b^2 > 4ac$
- no real roots: $b^2 < 4ac$

Example 14

Sketch the graph of $y = x^2 - 5x + 4$

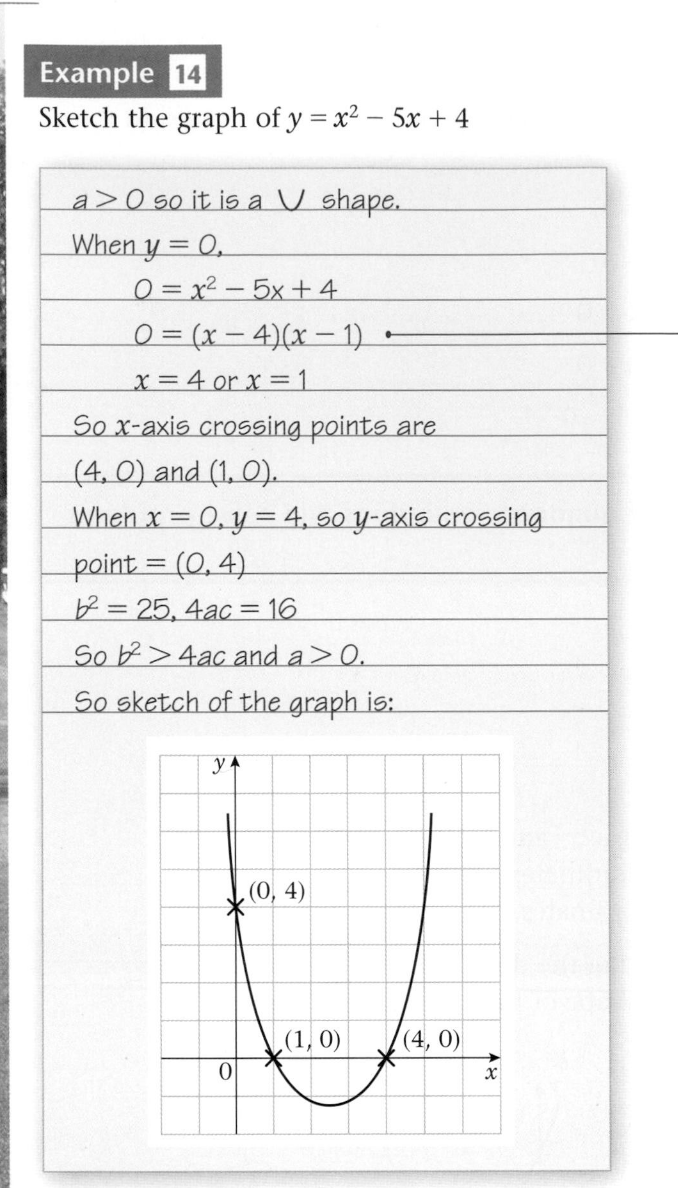

Factorise to solve the equation. (You may need to use the formula or complete the square.)

$a = 1$, $b = -5$, $c = 4$

Remember general shape:

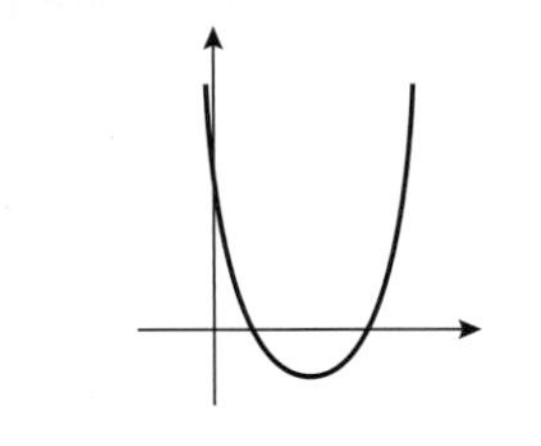

Label the crossing points.

Example 15

Find the values of k for which $x^2 + kx + 9 = 0$ has equal roots.

$$x^2 + kx + 9 = 0$$

Here $a = 1$, $b = k$ and $c = 9$

$$k^2 = 4 \times 1 \times 9$$

So $k = \pm 6$

For equal roots use $b^2 = 4ac$

Exercise 2F

1 Sketch the graphs of the following equations:

a $y = x^2 + 3x + 2$ **b** $y = x^2 - 3x + 10$ **c** $y = x^2 + 2x - 15$ **d** $y = 2x^2 + 7x + 3$

e $y = 2x^2 + x - 3$ **f** $y = 6x^2 - 19x + 10$ **g** $y = 3x^2 - 2x - 5$ **h** $y = 3x^2 - 13x$

i $y = -x^2 + 6x + 7$ **j** $y = 4 - 7x - 2x^2$

2 Find the values of k for which $x^2 + kx + 4 = 0$ has equal roots.

3 Find the values of k for which $kx^2 + 8x + k = 0$ has equal roots.

Mixed exercise 2G

1 Draw the graphs with the following equations, choosing appropriate values for x. For each graph write down the equation of the line of symmetry.

a $y = x^2 + 6x + 5$ **b** $y = 2x^2 - 3x - 4$

2 Solve the following equations:

a $y^2 + 3y + 2 = 0$ **b** $3x^2 + 13x - 10 = 0$

c $5x^2 - 10x = 4x + 3$ **d** $(2x - 5)^2 = 7$

3 Solve the following equations by:
i completing the square
ii using the formula.

a $x^2 + 5x + 2 = 0$ **b** $x^2 - 4x - 3 = 0$

c $5x^2 + 3x - 1 = 0$ **d** $3x^2 - 5x = 4$

4 Sketch graphs of the following equations:

a $y = x^2 + 5x + 4$ **b** $y = 2x^2 + x - 3$

c $y = 6 - 10x - 4x^2$ **d** $y = 15x - 2x^2$

5 Given that for all values of x:

$$3x^2 + 12x + 5 = p(x + q)^2 + r$$

a find the values of p, q and r

b solve the equation $3x^2 + 12x + 5 = 0$. *E*

6 Find, as surds, the roots of the equation:

$$2(x + 1)(x - 4) - (x - 2)^2 = 0$$

Hint: Remember roots mean solutions.

E

7 Use algebra to solve $(x - 1)(x + 2) = 18$.

Summary of key points

1 The general form of a quadratic equation is
$y = ax^2 + bx + c$ where a, b, c are constants and $a \neq 0$.

2 Quadratic equations can be solved by:

- factorisation
- completing the square:

$$x^2 + bx = \left(x + \frac{b}{2}\right)^2 - \left(\frac{b}{2}\right)^2$$

- using the formula

$$x = \frac{-b \pm \sqrt{(b^2 - 4ac)}}{2a}$$

3 A quadratic equation has two solutions, which may be equal.

4 To sketch a quadratic graph:

- decide on the shape:

  $a > 0$ ∪

  $a < 0$ ∩

- work out the x-axis and y-axis crossing points
- check the general shape by considering the discriminant $b^2 - 4ac$.

After completing this chapter you should be able to

1. solve simultaneous equations by elimination
2. solve simultaneous equations by substitution
3. solve linear and quadratic inequalities.

You will meet simultaneous equations on many occasions during the A level Mathematics course. In Core 1 you will use them to find where lines intersect. You will also use them to solve problems in sequences and series.

Equations and inequalities

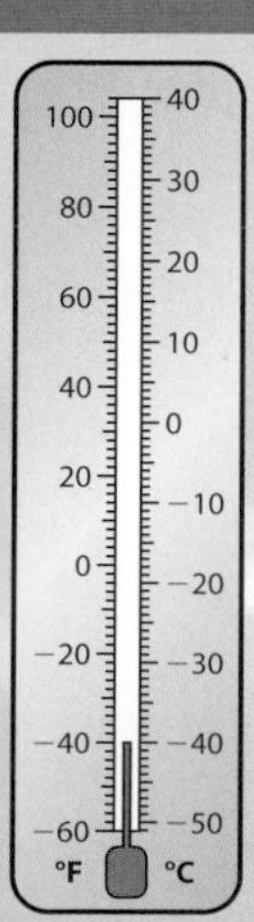

Did you know?

...that there is only one value at which the Fahrenheit and Celsius temperatures are the same?

Substitute F = C into F = 1.8C + 32. Solve the resulting linear equation. You should find that

$$C = F = -40$$

The temperature in the Arctic can reach **−40°F** which is identical to **−40°C**. So if you meet a polar bear and he asks you how cold it is you don't have to worry about units!

3.1 You can solve simultaneous linear equations by elimination.

Example 1

Solve the equations:

a $2x + 3y = 8$
$3x - y = 23$

b $4x - 5y = 4$
$6x + 2y = 25$

a $2x + 3y = 8$

$9x - 3y = 69$ — Multiply the 2nd equation by 3 to get $3y$ in each equation.

$11x = 77$ — Then add, since the $3y$ terms have different signs and y will be eliminated.

$x = 7$

First look for a way to eliminate x or y.

$14 + 3y = 8$ — Use $x = 7$ in the first equation to find y.

$3y = 8 - 14$

$y = -2$

So solution is $x = 7, y = -2$

You can consider the solution graphically.
The graph of each equation is a straight line.
The two straight lines intersect at $(7, -2)$.

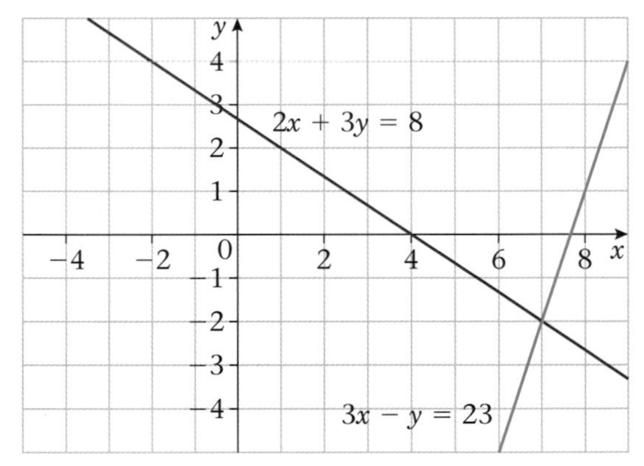

b $12x - 15y = 12$

$12x + 4y = 50$ — Multiply the first equation by 3 and multiply the 2nd equation by 2 to get $12x$ in each equation.

$-19y = -38$ — Subtract, since the $12x$ terms have the same sign (both positive).

$y = 2$

$4x - 10 = 4$ — Use $y = 2$ in the first equation to find the value of x.

$4x = 14$

$x = 3\frac{1}{2}$

So solution is $x = 3\frac{1}{2}, y = 2$

Graphically, each equation is a straight line.
The two straight lines intersect at (3.5, 2).

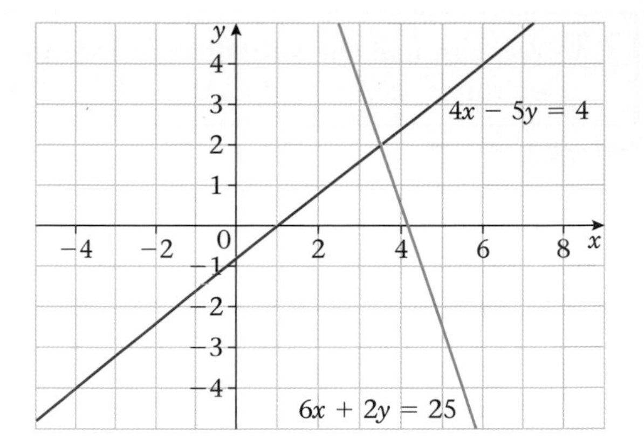

Exercise 3A

Solve these simultaneous equations by elimination:

1 $2x - y = 6$
$4x + 3y = 22$

2 $7x + 3y = 16$
$2x + 9y = 29$

3 $5x + 2y = 6$
$3x - 10y = 26$

4 $2x - y = 12$
$6x + 2y = 21$

5 $3x - 2y = -6$
$6x + 3y = 2$

6 $3x + 8y = 33$
$6x = 3 + 5y$

3.2 You can solve simultaneous linear equations by substitution.

Example 2

Solve the equations:

$2x - y = 1$
$4x + 2y = -30$

$y = 2x - 1$ — Rearrange an equation to get either $x = \ldots$ or $y = \ldots$ (here $y = \ldots$).

$4x + 2(2x - 1) = -30$ — Substitute this into the other equation (here in place of y).

$4x + 4x - 2 = -30$ — Solve for x.
$8x = -28$
$x = -3\frac{1}{2}$

$y = 2(-3\frac{1}{2}) - 1 = -8$ — Substitute $x = -3\frac{1}{2}$ into $y = 2x - 1$ to find the value of y.

So solution is $x = -3\frac{1}{2}$, $y = -8$.

Exercise 3B

Solve these simultaneous equations by substitution:

1 $x + 3y = 11$
$4x - 7y = 6$

2 $4x - 3y = 40$
$2x + y = 5$

3 $3x - y = 7$
$10x + 3y = -2$

4 $2y = 2x - 3$
$3y = x - 1$

3.3 You can use the substitution method to solve simultaneous equations where one equation is linear and the other is quadratic.

Example 3

Solve the equations:

a $x + 2y = 3$
$x^2 + 3xy = 10$

b $3x - 2y = 1$
$x^2 + y^2 = 25$

a $x = 3 - 2y$

Rearrange the linear equation to get $x = \dots$ or $y = \dots$ (here $x = \dots$).

$(3 - 2y)^2 + 3y(3 - 2y) = 10$

Substitute this into the quadratic equation (here in place of x).
$(3 - 2y)^2$ means $(3 - 2y)(3 - 2y)$ (see Chapter 1).

$9 - 12y + 4y^2 + 9y - 6y^2 = 10$

$-2y^2 - 3y - 1 = 0$

$2y^2 + 3y + 1 = 0$

$(2y + 1)(y + 1) = 0$

Solve for y using factorisation.

$y = -\frac{1}{2}$ or $y = -1$

So $x = 4$ or $x = 5$

Find the corresponding x-values by substituting the y-values into $x = 3 - 2y$.

Solutions are $x = 4, y = -\frac{1}{2}$

and $x = 5, y = -1$

There are two solution pairs. The graph of the linear equation (straight line) would intersect the graph of the quadratic (curve) at two points.

b $3x - 2y = 1$

$2y = 3x - 1$

$y = \frac{3x - 1}{2}$

Find $y = \dots$ from linear equation.

$x^2 + \left(\frac{3x - 1}{2}\right)^2 = 25$

Substitute $y = \frac{3x - 1}{2}$ into the quadratic equation to form an equation in x.

$x^2 + \left(\frac{9x^2 - 6x + 1}{4}\right) = 25$

Now multiply by 4.

$4x^2 + 9x^2 - 6x + 1 = 100$

$13x^2 - 6x - 99 = 0$

$(13x + 33)(x - 3) = 0$

$x = -\frac{33}{13}$ or $x = 3$

Solve for x.

$y = -\frac{56}{13}$ or $y = 4$

Substitute x-values into $y = \frac{3x - 1}{2}$.

Solutions are $x = 3, y = 4$

and $x = -\frac{33}{13}, y = -\frac{56}{13}$

Graphically, the linear equation (straight line) intersects the quadratic equation (curve) at two points.

(This curve is a circle. You will learn about its equation in Book C2.)

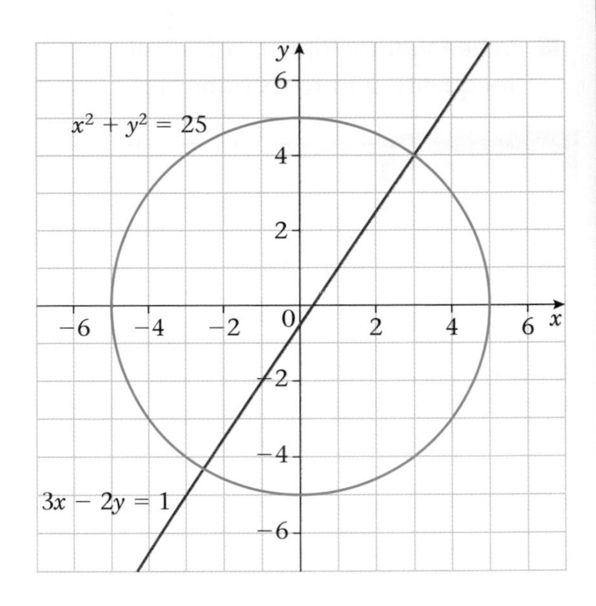

It is possible, of course, that a given straight line and a given curve do *not* intersect. In this case, the quadratic equation that has to be solved would have no real roots (in this case $b^2 - 4ac < 0$). (See Section 2.6.)

Exercise 3C

1 Solve the simultaneous equations:

a $x + y = 11$
$xy = 30$

b $2x + y = 1$
$x^2 + y^2 = 1$

c $y = 3x$
$2y^2 - xy = 15$

d $x + y = 9$
$x^2 - 3xy + 2y^2 = 0$

e $3a + b = 8$
$3a^2 + b^2 = 28$

f $2u + v = 7$
$uv = 6$

2 Find the coordinates of the points at which the line with equation $y = x - 4$ intersects the curve with equation $y^2 = 2x^2 - 17$.

3 Find the coordinates of the points at which the line with equation $y = 3x - 1$ intersects the curve with equation $y^2 - xy = 15$.

4 Solve the simultaneous equations:

a $3x + 2y = 7$
$x^2 + y = 8$

b $2x + 2y = 7$
$x^2 - 4y^2 = 8$

5 Solve the simultaneous equations, giving your answers in their simplest surd form:

a $x - y = 6$
$xy = 4$

b $2x + 3y = 13$
$x^2 + y^2 = 78$

3.4 You can solve linear inequalities using similar methods to those for solving linear equations.

You need to be careful when you multiply or divide an inequality by a negative number. You need to turn round the inequality sign:

$$5 > 2$$

Multiply by -2 $\quad -10 < -4$

■ **When you multiply or divide an inequality by a negative number, you need to change the inequality sign to its opposite.**

Example 4

Find the set of values of x for which:

a $2x - 5 < 7$

b $5x + 9 \geqslant x + 20$

c $12 - 3x < 27$

d $3(x - 5) > 5 - 2(x - 8)$

a $2x - 5 < 7$

$2x < 12$ — Add 5 to both sides.

$x < 6$ — Divide both sides by 2.

b $5x + 9 \geqslant x + 20$

$4x + 9 \geqslant 20$ — Subtract x from both sides.

$4x \geqslant 11$ — Subtract 9 from both sides.

$x \geqslant 2.75$ — Divide both sides by 4.

For **c**, two approaches are shown:

c $12 - 3x < 27$

$-3x < 15$ — Subtract 12 from both sides.

$x > -5$ — Divide both sides by -3. (You therefore need to turn round the inequality sign.)

$12 - 3x < 27$

$12 < 27 + 3x$ — Add $3x$ to both sides.

$-15 < 3x$ — Subtract 27 from both sides.

$-5 < x$ — Divide both sides by 3.

$x > -5$ — Rewrite with x on LHS.

d $3(x - 5) > 5 - 2(x - 8)$

$3x - 15 > 5 - 2x + 16$ — Multiply out (note: $-2 \times -8 = +16$).

$5x > 5 + 16 + 15$ — Add 15 to both sides.

$5x > 36$

$x > 7.2$ — Divide both sides by 5.

You may sometimes need to find the set of values of x for which two inequalities are true together. Number lines are helpful here.

Example 5

Find the set of values of x for which:

$$3x - 5 < x + 8 \text{ and } 5x > x - 8$$

$3x - 5 < x + 8$	$5x > x - 8$
$2x - 5 < 8$	$4x > -8$
$2x < 13$	$x > -2$
$x < 6.5$	

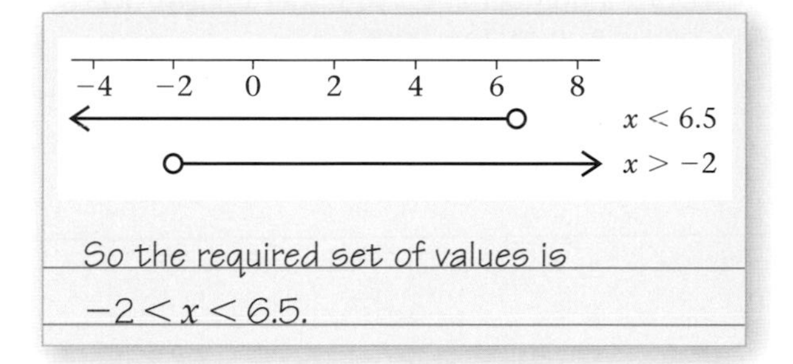

So the required set of values is $-2 < x < 6.5$.

Draw a number line to illustrate the two inequalities.

The 'hollow dots' at the end of each line show that the end value is not included in the set of values.

Show an included end value (⩽ or ⩾) by using a 'solid dot' (●).

The two sets of values overlap (or intersect) where $-2 < x < 6.5$.

Notice here how this is written when x lies between two values.

Example 6

Find the set of values of x for which:

$$x - 5 > 1 - x \text{ and } 15 - 3x > 5 + 2x$$

$x - 5 > 1 - x$	$15 - 3x > 5 + 2x$
$2x - 5 > 1$	$10 - 3x > 2x$
$2x > 6$	$10 > 5x$
$x > 3$	$2 > x$
	$x < 2$

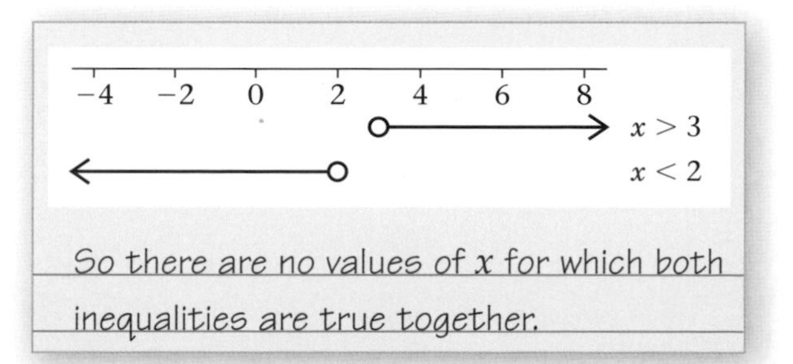

So there are no values of x for which both inequalities are true together.

Draw a number line. Note that there is no overlap between the two sets of values.

Example 7

Find the set of values of x for which:

$4x + 7 > 3$ and $17 < 11 + 2x$

$4x + 7 > 3$	$17 < 11 + 2x$
$4x > -4$	$17 - 11 < 2x$
$x > -1$	$6 < 2x$
	$3 < x$
	$x > 3$

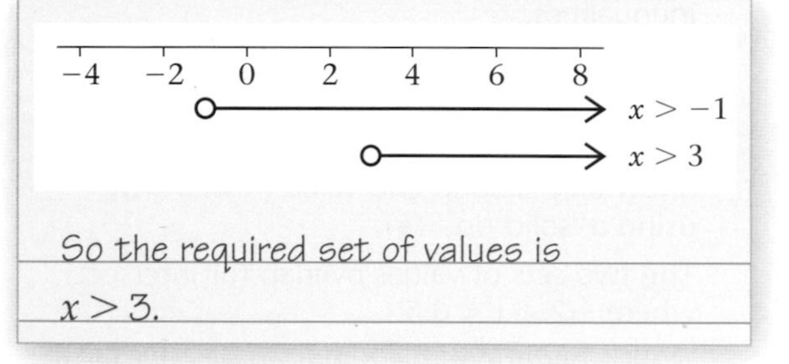

Draw a number line. Note that the two sets of values overlap where $x > 3$.

So the required set of values is $x > 3$.

Exercise 3D

1 Find the set of values of x for which:

a $2x - 3 < 5$ **b** $5x + 4 \geqslant 39$

c $6x - 3 > 2x + 7$ **d** $5x + 6 \leqslant -12 - x$

e $15 - x > 4$ **f** $21 - 2x > 8 + 3x$

g $1 + x < 25 + 3x$ **h** $7x - 7 < 7 - 7x$

i $5 - 0.5x \geqslant 1$ **j** $5x + 4 > 12 - 2x$

2 Find the set of values of x for which:

a $2(x - 3) \geqslant 0$ **b** $8(1 - x) > x - 1$

c $3(x + 7) \leqslant 8 - x$ **d** $2(x - 3) - (x + 12) < 0$

e $1 + 11(2 - x) < 10(x - 4)$ **f** $2(x - 5) \geqslant 3(4 - x)$

g $12x - 3(x - 3) < 45$ **h** $x - 2(5 + 2x) < 11$

i $x(x - 4) \geqslant x^2 + 2$ **j** $x(5 - x) \geqslant 3 + x - x^2$

3 Find the set of values of x for which:

a $3(x - 2) > x - 4$ and $4x + 12 > 2x + 17$

b $2x - 5 < x - 1$ and $7(x + 1) > 23 - x$

c $2x - 3 > 2$ and $3(x + 2) < 12 + x$

d $15 - x < 2(11 - x)$ and $5(3x - 1) > 12x + 19$

e $3x + 8 \leqslant 20$ and $2(3x - 7) \geqslant x + 6$

3.5 To solve a quadratic inequality you
- **solve the corresponding quadratic equation, then**
- **sketch the graph of the quadratic function, then**
- **use your sketch to find the required set of values.**

Example 8

Find the set of values of x for which $x^2 - 4x - 5 < 0$ and draw a sketch to show this.

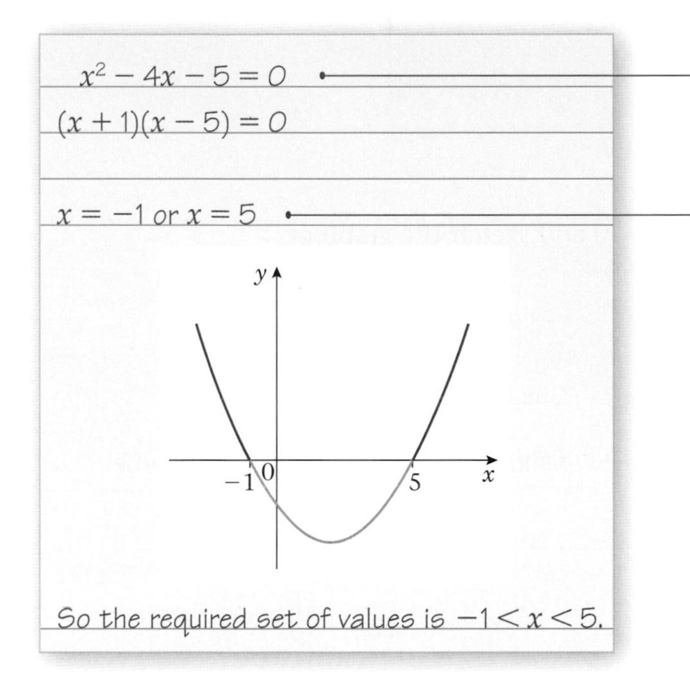

Quadratic equation.

Factorise (or use the quadratic formula). (See Section 2.5.)

−1 and 5 are called critical values.

Your sketch does not need to be accurate. All you really need to know is that the graph is '⋃-shaped' and crosses the x-axis at −1 and 5. (See Section 2.6.)

$x^2 - 4x - 5 < 0$ ($y < 0$) for the part of the graph below the x-axis, as shown by the paler part in the rough sketch.

Example 9

Find the set of values of x for which $x^2 - 4x - 5 > 0$.

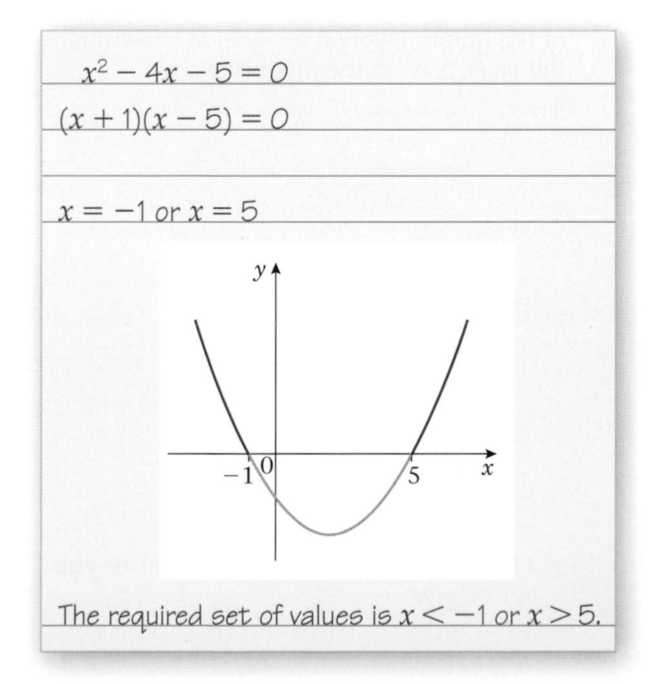

The only difference between this example and the previous example is that it has to be greater than 0 (> 0). The solution would be exactly the same apart from the final stage.

$x^2 - 4x - 5 > 0$ ($y > 0$) for the part of the graph above the x-axis, as shown by the darker parts of the rough sketch in Example 8.

Be careful how you write down solutions like those on page 33.

$-1 < x < 5$ is fine, showing that x is between -1 and 5.

But it is wrong to write something like $5 < x < -1$ or $-1 > x > 5$ because x cannot be less than -1 and greater than 5 at the same time.

This type of solution (the darker parts of the graph) needs to be written in two separate parts, $x < -1$, $x > 5$.

Example 10

Find the set of values of x for which $3 - 5x - 2x^2 < 0$ and sketch the graph of $y = 3 - 5x - 2x^2$.

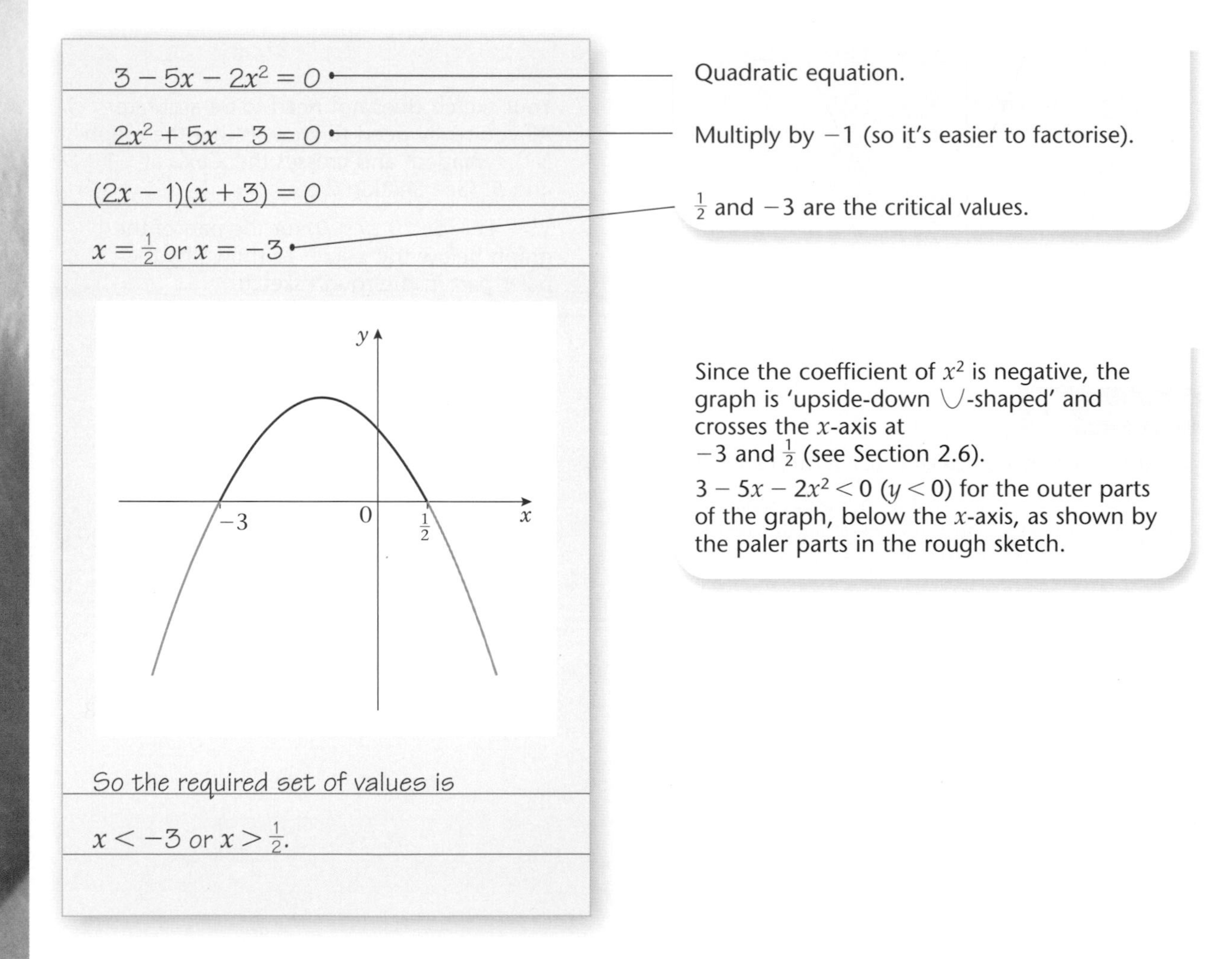

You may have to rearrange the quadratic inequality to get all the terms 'on one side' before you can solve it, as shown in the next example.

Example 11

Find the set of values of x for which $12 + 4x > x^2$.

Method 1: sketch graph

There are two possible approaches for Method 1, depending on which side of the inequality sign you put the expression.

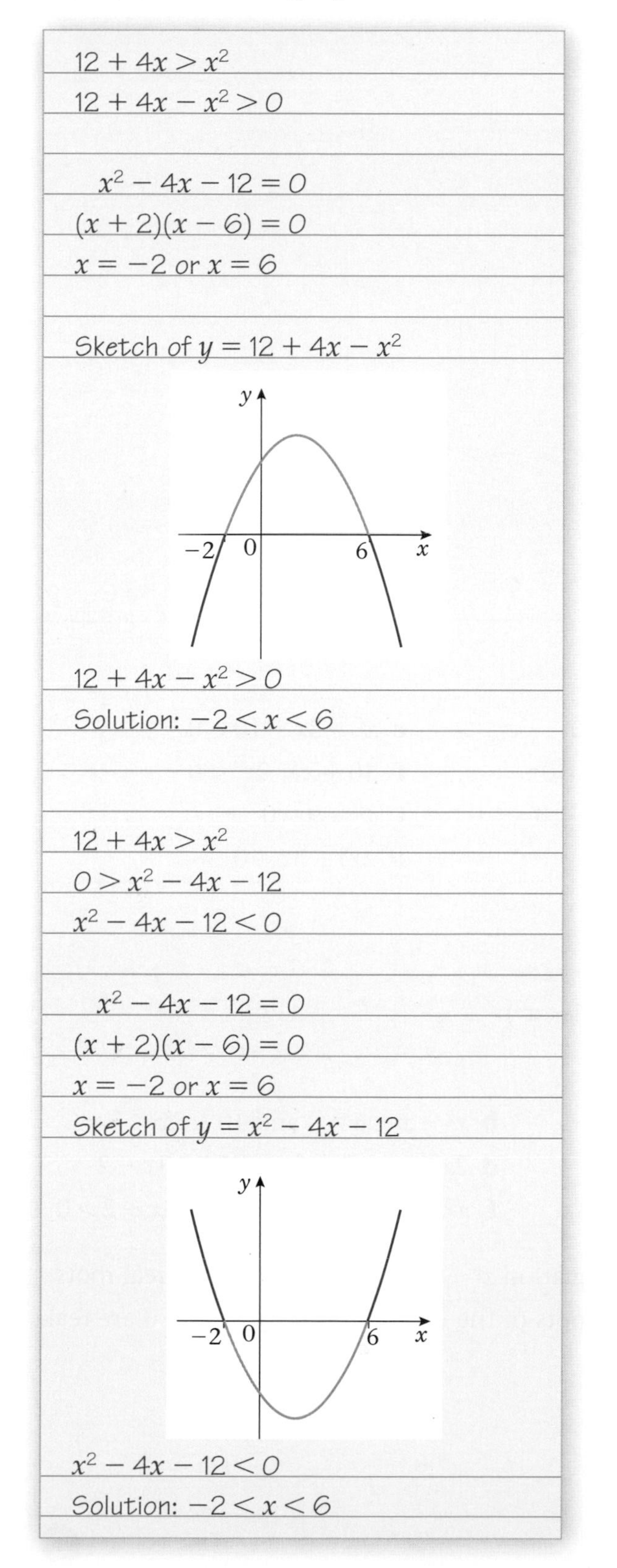

$12 + 4x > x^2$

$12 + 4x - x^2 > 0$

$x^2 - 4x - 12 = 0$

$(x + 2)(x - 6) = 0$

$x = -2$ or $x = 6$

Sketch of $y = 12 + 4x - x^2$

$12 + 4x - x^2 > 0$

Solution: $-2 < x < 6$

$12 + 4x > x^2$

$0 > x^2 - 4x - 12$

$x^2 - 4x - 12 < 0$

$x^2 - 4x - 12 = 0$

$(x + 2)(x - 6) = 0$

$x = -2$ or $x = 6$

Sketch of $y = x^2 - 4x - 12$

$x^2 - 4x - 12 < 0$

Solution: $-2 < x < 6$

Find the set of values of x for which $12 + 4x > x^2$.

Method 2: table

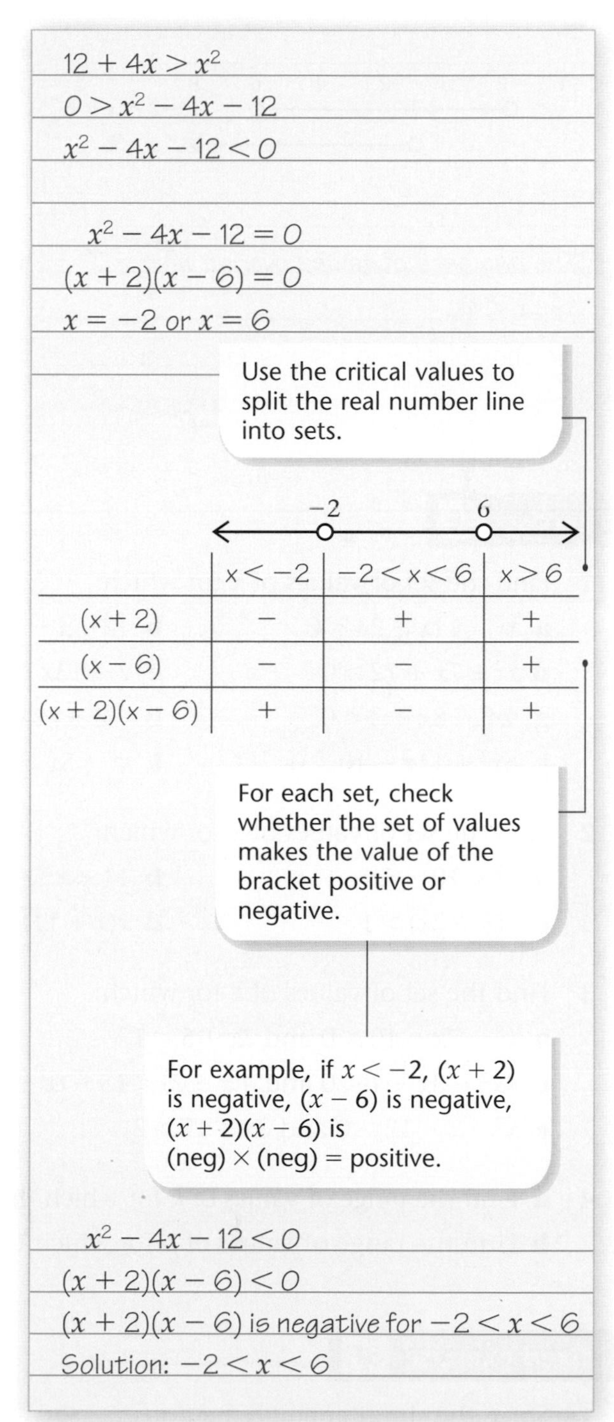

$12 + 4x > x^2$

$0 > x^2 - 4x - 12$

$x^2 - 4x - 12 < 0$

$x^2 - 4x - 12 = 0$

$(x + 2)(x - 6) = 0$

$x = -2$ or $x = 6$

Use the critical values to split the real number line into sets.

	$x < -2$	$-2 < x < 6$	$x > 6$
$(x + 2)$	−	+	+
$(x - 6)$	−	−	+
$(x + 2)(x - 6)$	+	−	+

For each set, check whether the set of values makes the value of the bracket positive or negative.

For example, if $x < -2$, $(x + 2)$ is negative, $(x - 6)$ is negative, $(x + 2)(x - 6)$ is (neg) × (neg) = positive.

$x^2 - 4x - 12 < 0$

$(x + 2)(x - 6) < 0$

$(x + 2)(x - 6)$ is negative for $-2 < x < 6$

Solution: $-2 < x < 6$

Example 12

Find the set of values of x for which $12 + 4x > x^2$ and $5x - 3 > 2$.

Solving $12 + 4x > x^2$ gives $-2 < x < 6$ (see Example 11).

Solving $5x - 3 > 2$ gives $x > 1$.

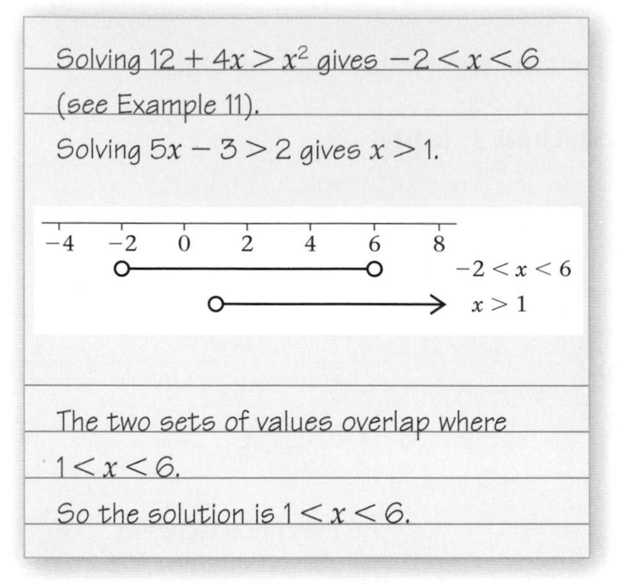

The two sets of values overlap where $1 < x < 6$.

So the solution is $1 < x < 6$.

Exercise 3E

1 Find the set of values of x for which:

a $x^2 - 11x + 24 < 0$
b $12 - x - x^2 > 0$
c $x^2 - 3x - 10 > 0$
d $x^2 + 7x + 12 \geqslant 0$
e $7 + 13x - 2x^2 > 0$
f $10 + x - 2x^2 < 0$
g $4x^2 - 8x + 3 \leqslant 0$
h $-2 + 7x - 3x^2 < 0$
i $x^2 - 9 < 0$
j $6x^2 + 11x - 10 > 0$
k $x^2 - 5x > 0$
l $2x^2 + 3x \leqslant 0$

2 Find the set of values of x for which:

a $x^2 < 10 - 3x$
b $11 < x^2 + 10$
c $x(3 - 2x) > 1$
d $x(x + 11) < 3(1 - x^2)$

3 Find the set of values of x for which:

a $x^2 - 7x + 10 < 0$ and $3x + 5 < 17$
b $x^2 - x - 6 > 0$ and $10 - 2x < 5$
c $4x^2 - 3x - 1 < 0$ and $4(x + 2) < 15 - (x + 7)$
d $2x^2 - x - 1 < 0$ and $14 < 3x - 2$
e $x^2 - x - 12 > 0$ and $3x + 17 > 2$
f $x^2 - 2x - 3 < 0$ and $x^2 - 3x + 2 > 0$

4 **a** Find the range of values of k for which the equation $x^2 - kx + (k + 3) = 0$ has no real roots.
b Find the range of values of p for which the roots of the equation $px^2 + px - 2 = 0$ are real.

Mixed exercise 3F

1 Solve the simultaneous equations:

$$x + 2y = 3$$
$$x^2 - 4y^2 = -33$$

E

2 Show that the elimination of x from the simultaneous equations

$$x - 2y = 1$$
$$3xy - y^2 = 8$$

produces the equation

$$5y^2 + 3y - 8 = 0.$$

Solve this quadratic equation and hence find the pairs (x, y) for which the simultaneous equations are satisfied. *E*

3 **a** Given that $3^x = 9^{y-1}$, show that $x = 2y - 2$.

b Solve the simultaneous equations:

$$x = 2y - 2$$
$$x^2 = y^2 + 7$$

E

4 Solve the simultaneous equations:

$$x + 2y = 3$$
$$x^2 - 2y + 4y^2 = 18$$

E

5 **a** Solve the inequality $3x - 8 > x + 13$.

b Solve the inequality $x^2 - 5x - 14 > 0$. *E*

6 Find the set of values of x for which $(x - 1)(x - 4) < 2(x - 4)$. *E*

7 **a** Use algebra to solve $(x - 1)(x + 2) = 18$.

b Hence, or otherwise, find the set of values of x for which $(x - 1)(x + 2) > 18$. *E*

8 Find the set of values of x for which:

a $6x - 7 < 2x + 3$

b $2x^2 - 11x + 5 < 0$

c both $6x - 7 < 2x + 3$ and $2x^2 - 11x + 5 < 0$. *E*

9 Find the values of k for which $kx^2 + 8x + 5 = 0$ has real roots.

10 Find algebraically the set of values of x for which $(2x - 3)(x + 2) > 3(x - 2)$. *E*

11 **a** Find, as surds, the roots of the equation $2(x + 1)(x - 4) - (x - 2)^2 = 0$.

b Hence find the set of values of x for which $2(x + 1)(x - 4) - (x - 2)^2 > 0$. *E*

12 **a** Use algebra to find the set of values of x for which $x(x - 5) > 36$.

b Using your answer to part **a**, find the set of values of y for which $y^2(y^2 - 5) > 36$.

13 The specification for a rectangular car park states that the length x m is to be 5 m more than the breadth. The perimeter of the car park is to be greater than 32 m. *E*

a Form a linear inequality in x.

The area of the car park is to be less than 104 m^2.

b Form a quadratic inequality in x.

c By solving your inequalities, determine the set of possible values of x. *E*

Summary of key points

1 You can solve linear simultaneous equations by elimination or substitution.

2 You can use the substitution method to solve simultaneous equations, where one equation is linear and the other is quadratic. You usually start by finding an expression for x or y from the linear equation.

3 When you multiply or divide an inequality by a negative number, you need to change the inequality sign to its opposite.

4 To solve a quadratic inequality you
- solve the corresponding quadratic equation, then
- sketch the graph of the quadratic function, then
- use your sketch to find the required set of values.

After completing this chapter you should be able to

1. sketch cubic graphs
2. sketch the graph of the reciprocal function $y = \frac{k}{x}$
3. find where curves intersect
4. understand how the transformations $f(x + a)$, $f(x) + a$, $f(ax)$ and $af(x)$ affect the graph of the curve $y = f(x)$.

You will analyse graphs in greater detail when you start differentiation. It is worth remembering the techniques in this chapter, because they will provide further information about the shape of the function. Later on in the course you will be asked to sketch complex graphs which are simple transformations of a standard function..

4 Sketching curves

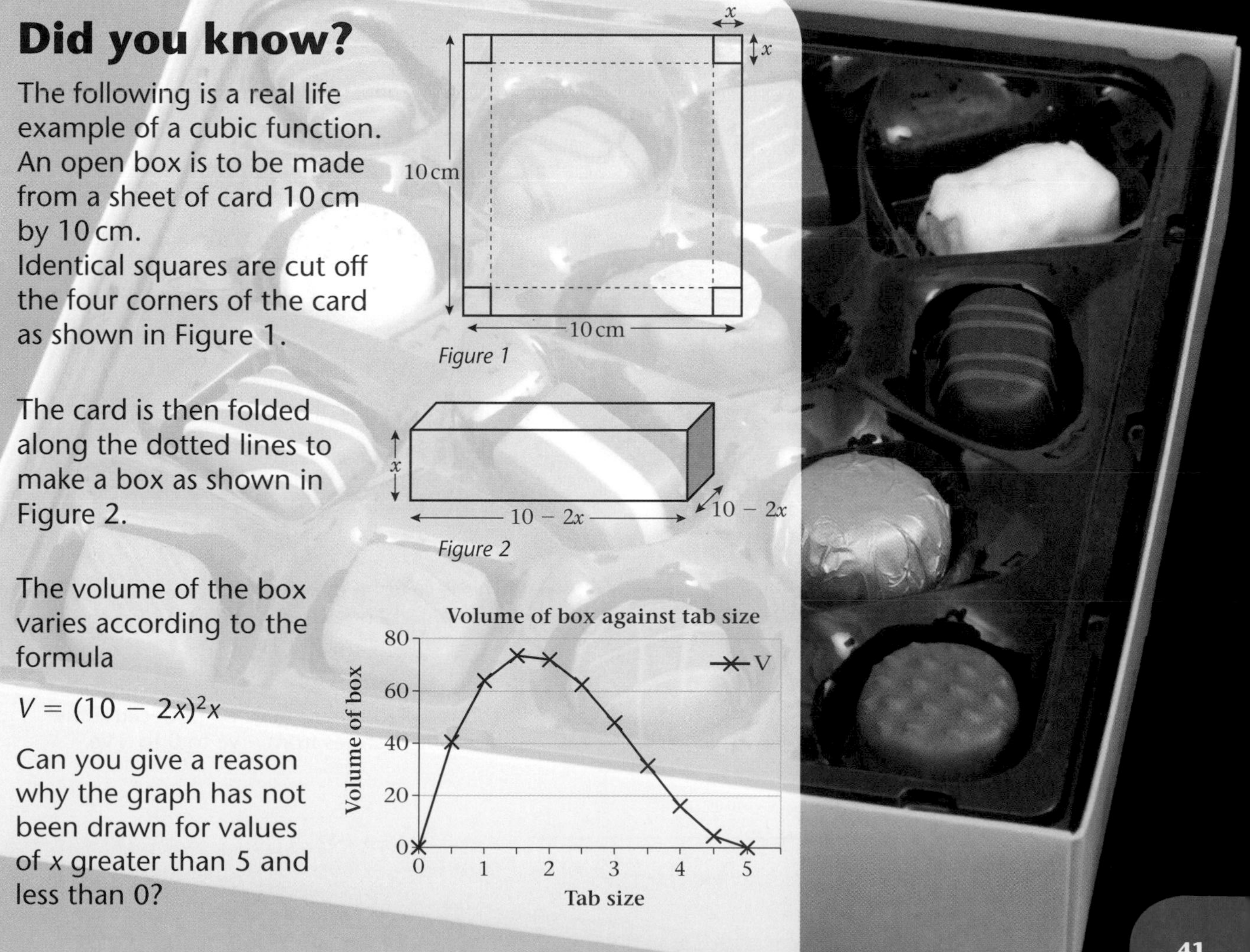

Did you know?

The following is a real life example of a cubic function. An open box is to be made from a sheet of card 10 cm by 10 cm.
Identical squares are cut off the four corners of the card as shown in Figure 1.

Figure 1

The card is then folded along the dotted lines to make a box as shown in Figure 2.

Figure 2

The volume of the box varies according to the formula

$V = (10 - 2x)^2x$

Can you give a reason why the graph has not been drawn for values of x greater than 5 and less than 0?

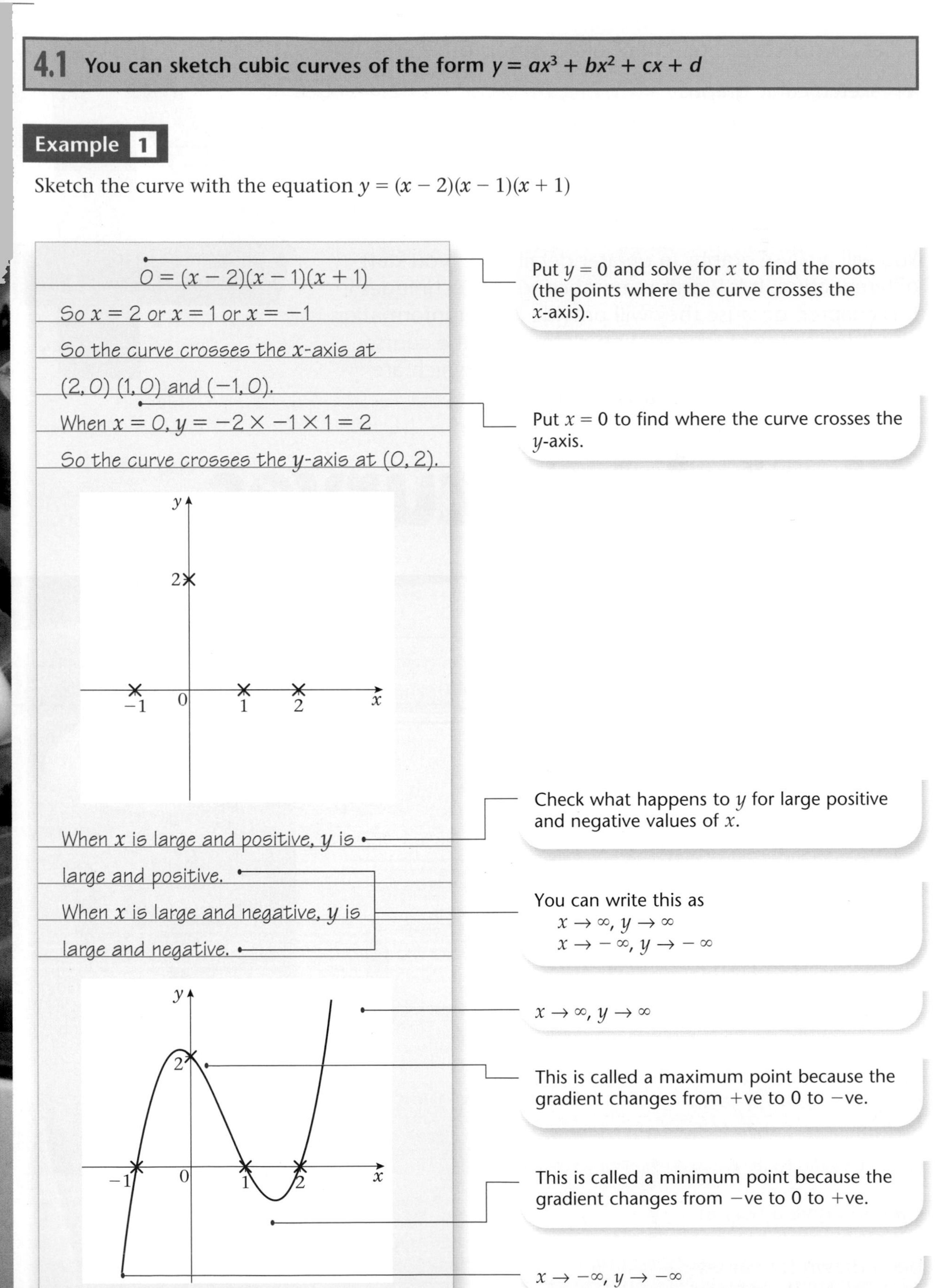
4.1 You can sketch cubic curves of the form $y = ax^3 + bx^2 + cx + d$
Example 1
Sketch the curve with the equation $y = (x - 2)(x - 1)(x + 1)$
$0 = (x - 2)(x - 1)(x + 1)$
So $x = 2$ or $x = 1$ or $x = -1$
So the curve crosses the x-axis at
(2, 0) (1, 0) and (−1, 0).
When $x = 0$, $y = -2 \times -1 \times 1 = 2$
So the curve crosses the y-axis at (0, 2).
Put $y = 0$ and solve for x to find the roots (the points where the curve crosses the x-axis).
Put $x = 0$ to find where the curve crosses the y-axis.
y
2
−1
0
1
2
x
When x is large and positive, y is large and positive.
When x is large and negative, y is large and negative.
Check what happens to y for large positive and negative values of x.
You can write this as
$x \to \infty, y \to \infty$
$x \to -\infty, y \to -\infty$
y
2
−1
0
1
2
x
$x \to \infty, y \to \infty$
This is called a maximum point because the gradient changes from +ve to 0 to −ve.
This is called a minimum point because the gradient changes from −ve to 0 to +ve.
$x \to -\infty, y \to -\infty$

In your exam you will not be expected to work out the coordinates of the maximum or minimum points without further work, but you should mark points where the curve meets the axes.

Example 2

Sketch the curves with the following equations and show the points where they cross the coordinate axes.

a $y = (x - 2)(1 - x)(1 + x)$ **b** $y = x(x + 1)(x + 2)$

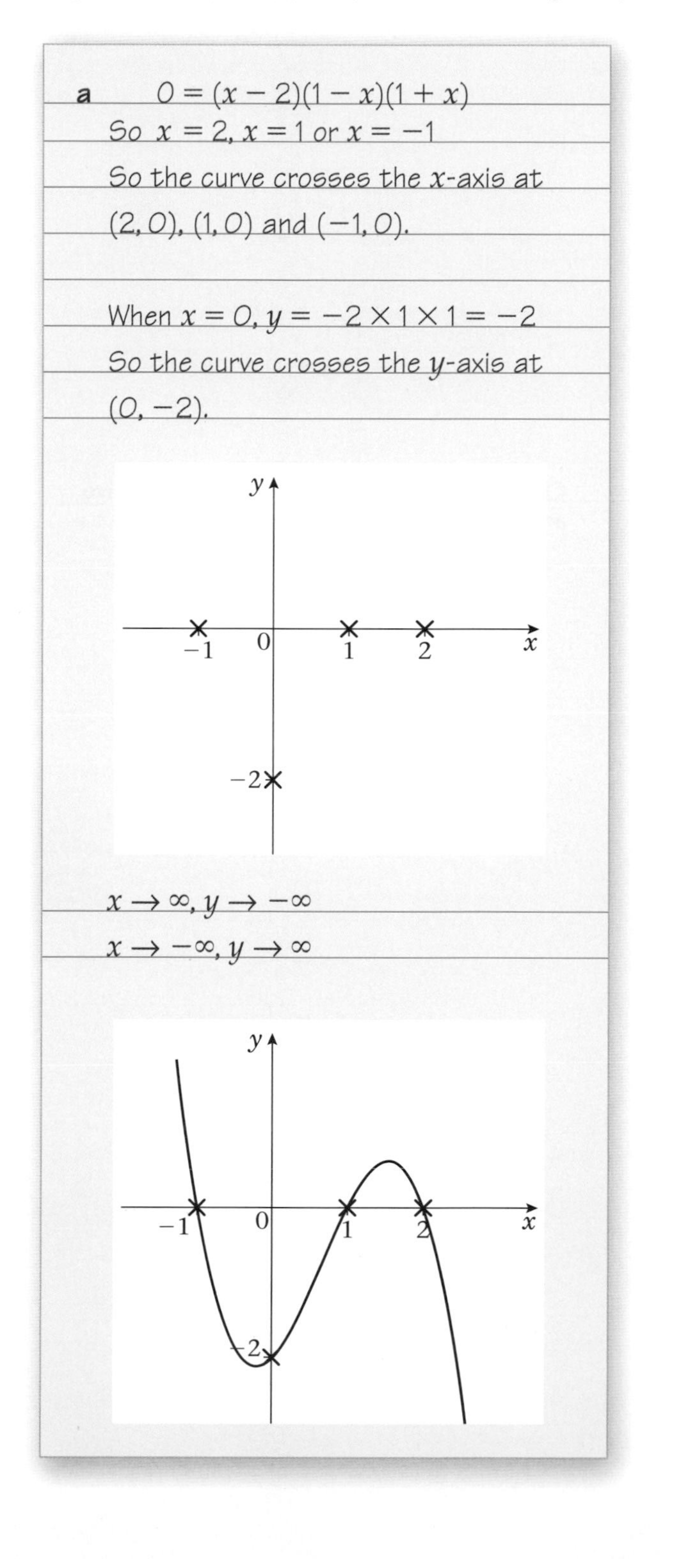

Put $y = 0$ and solve for x.

Find the value of y when $x = 0$.

Check what happens to y for large positive and negative values of x.

Notice that this curve is a reflection in the x-axis of the curve in Example 1.

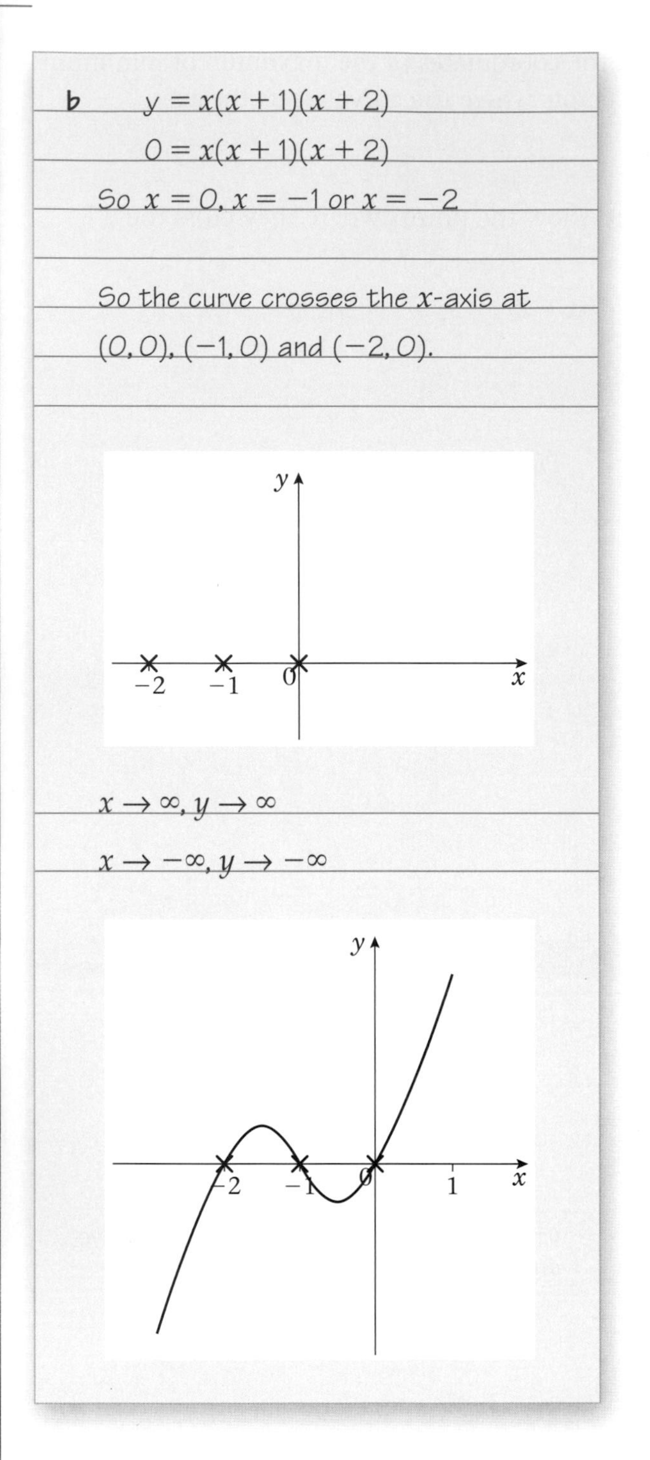

Put $y = 0$ and solve for x.

So the curve crosses the y-axis at (0, 0).

Check what happens to y for large positive and negative values of x.

Example 3

Sketch the following curves.

a $y = (x - 1)^2(x + 1)$

b $y = x^3 - 2x^2 - 3x$

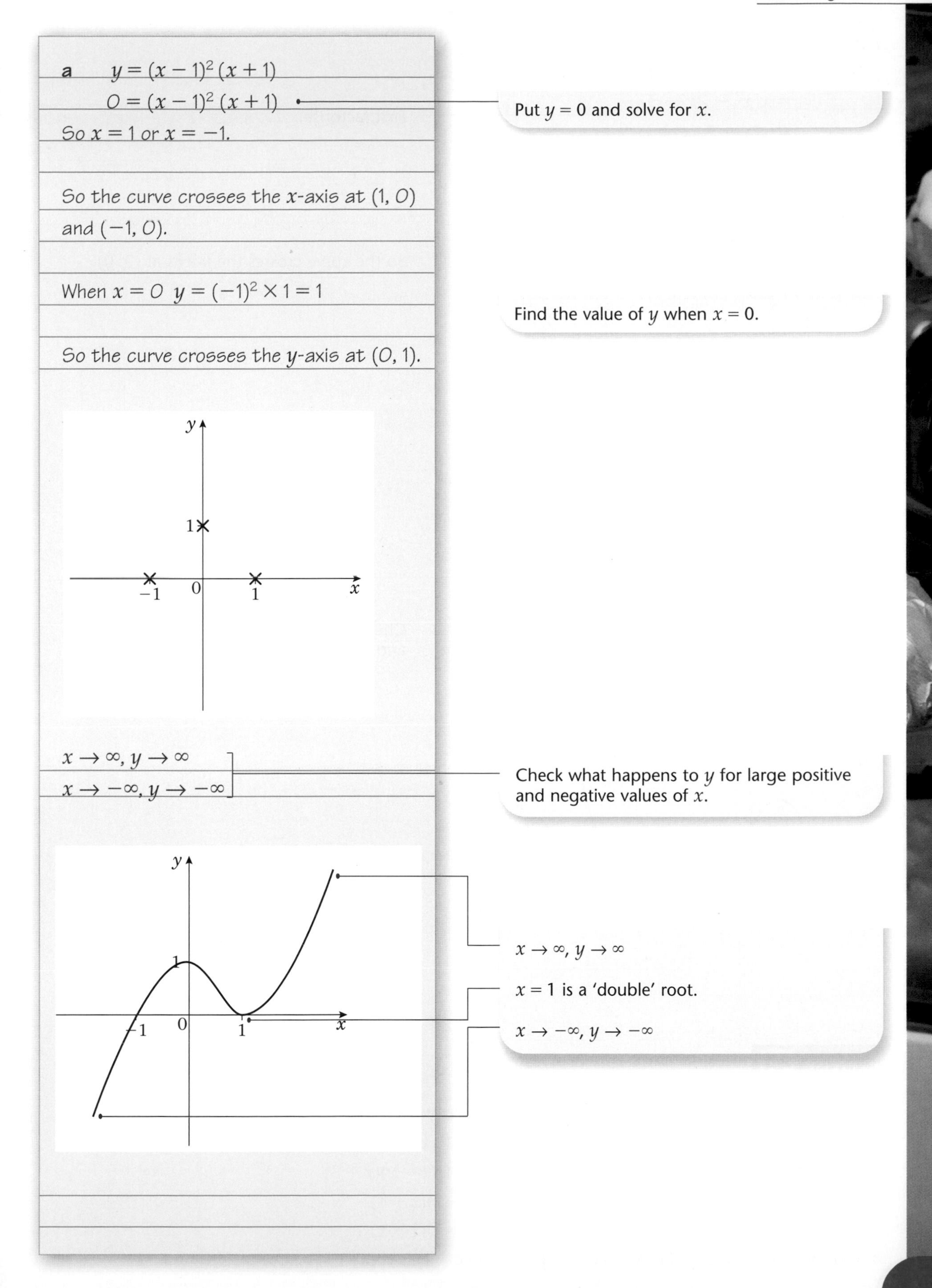

a $y = (x-1)^2(x+1)$

$0 = (x-1)^2(x+1)$

So $x = 1$ or $x = -1$.

Put $y = 0$ and solve for x.

So the curve crosses the x-axis at $(1, 0)$ and $(-1, 0)$.

When $x = 0$ $y = (-1)^2 \times 1 = 1$

Find the value of y when $x = 0$.

So the curve crosses the y-axis at $(0, 1)$.

$x \to \infty, y \to \infty$

$x \to -\infty, y \to -\infty$

Check what happens to y for large positive and negative values of x.

$x \to \infty, y \to \infty$

$x = 1$ is a 'double' root.

$x \to -\infty, y \to -\infty$

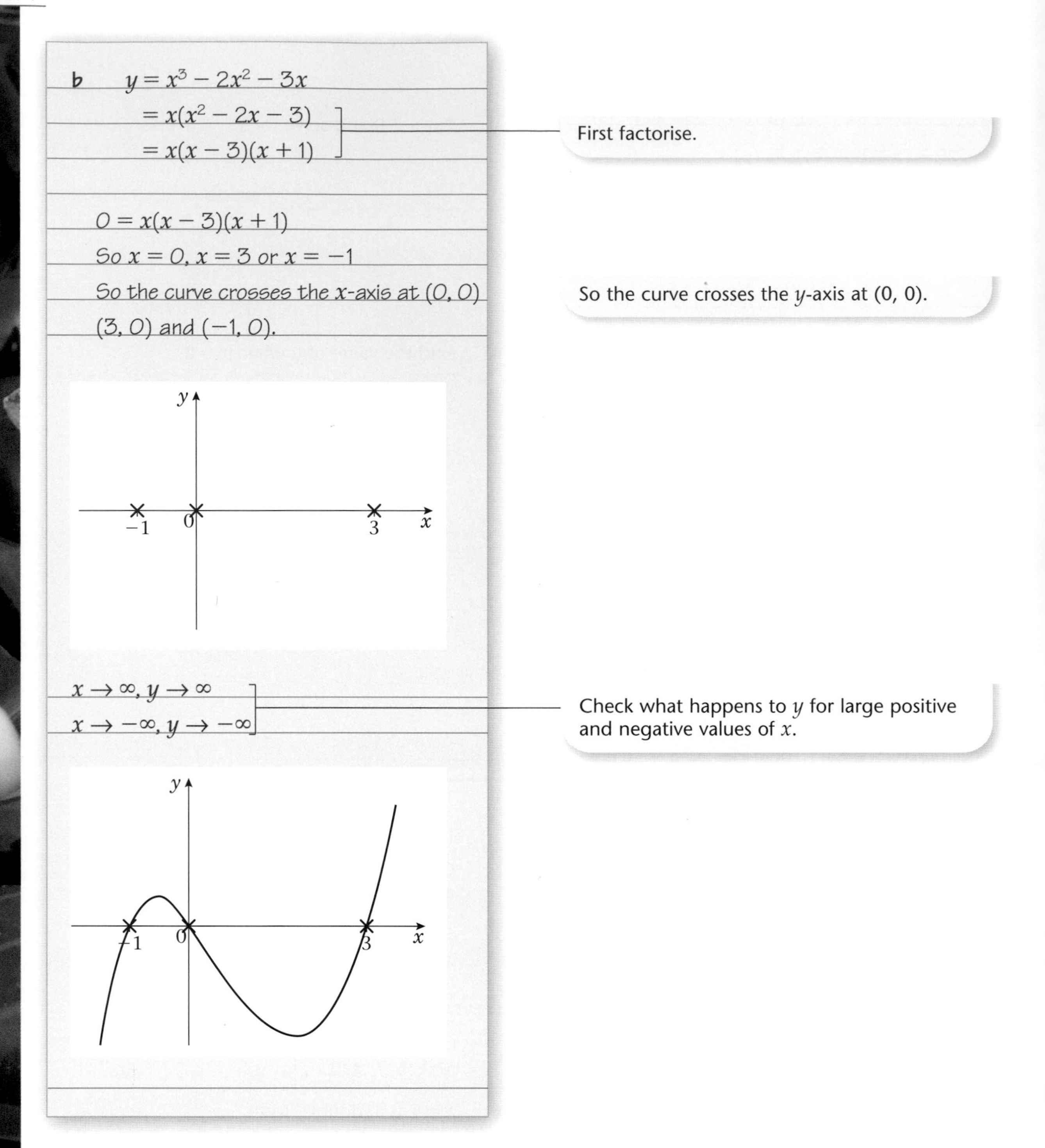

b $y = x^3 - 2x^2 - 3x$

$= x(x^2 - 2x - 3)$

$= x(x - 3)(x + 1)$

First factorise.

$0 = x(x - 3)(x + 1)$

So $x = 0$, $x = 3$ or $x = -1$

So the curve crosses the x-axis at $(0, 0)$ $(3, 0)$ and $(-1, 0)$.

So the curve crosses the y-axis at (0, 0).

$x \to \infty, y \to \infty$

$x \to -\infty, y \to -\infty$

Check what happens to y for large positive and negative values of x.

Exercise 4A

1 Sketch the following curves and indicate clearly the points of intersection with the axes:

a $y = (x - 3)(x - 2)(x + 1)$ **b** $y = (x - 1)(x + 2)(x + 3)$

c $y = (x + 1)(x + 2)(x + 3)$ **d** $y = (x + 1)(1 - x)(x + 3)$

e $y = (x - 2)(x - 3)(4 - x)$ **f** $y = x(x - 2)(x + 1)$

g $y = x(x + 1)(x - 1)$ **h** $y = x(x + 1)(1 - x)$

i $y = (x - 2)(2x - 1)(2x + 1)$ **j** $y = x(2x - 1)(x + 3)$

2 Sketch the curves with the following equations:

a $y = (x + 1)^2(x - 1)$ **b** $y = (x + 2)(x - 1)^2$

c $y = (2 - x)(x + 1)^2$ **d** $y = (x - 2)(x + 1)^2$

e $y = x^2(x + 2)$ **f** $y = (x - 1)^2x$

g $y = (1 - x)^2(3 + x)$ **h** $y = (x - 1)^2(3 - x)$

i $y = x^2(2 - x)$ **j** $y = x^2(x - 2)$

3 Factorise the following equations and then sketch the curves:

a $y = x^3 + x^2 - 2x$ **b** $y = x^3 + 5x^2 + 4x$

c $y = x^3 + 2x^2 + x$ **d** $y = 3x + 2x^2 - x^3$

e $y = x^3 - x^2$ **f** $y = x - x^3$

g $y = 12x^3 - 3x$ **h** $y = x^3 - x^2 - 2x$

i $y = x^3 - 9x$ **j** $y = x^3 - 9x^2$

4.2 You need to be able to sketch and interpret graphs of cubic functions of the form $y = x^3$.

Example 4

Sketch the curve with equation $y = x^3$.

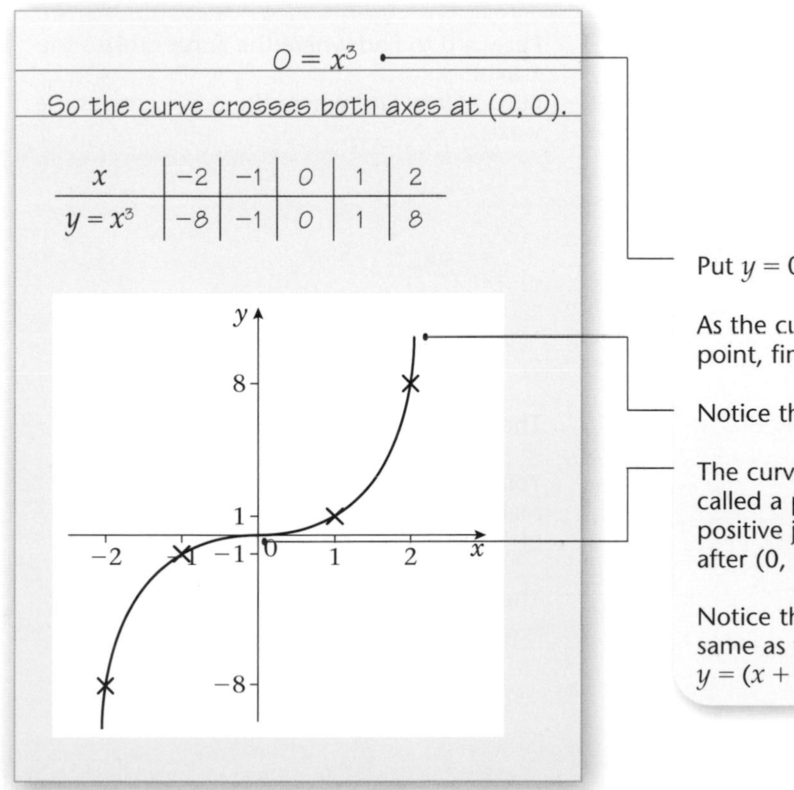

$0 = x^3$

So the curve crosses both axes at (0, 0).

x	−2	−1	0	1	2
$y = x^3$	−8	−1	0	1	8

Put $y = 0$ and solve for x.

As the curve passes the axes at only one point, find its shape by plotting a few points.

Notice that as x increases, y increases rapidly.

The curve is 'flat' at (0, 0). This point is called a point of inflexion. The gradient is positive just before (0, 0) and positive just after (0, 0).

Notice that the shape of this curve is the same as the curve with equation $y = (x + 1)^3$, which is shown in Example 5.

Example 5

Sketch the curve with equations:

a $y = -x^3$ **b** $y = (x + 1)^3$ **c** $y = (3 - x)^3$

Show their positions relative to the curve with equation $y = x^3$.

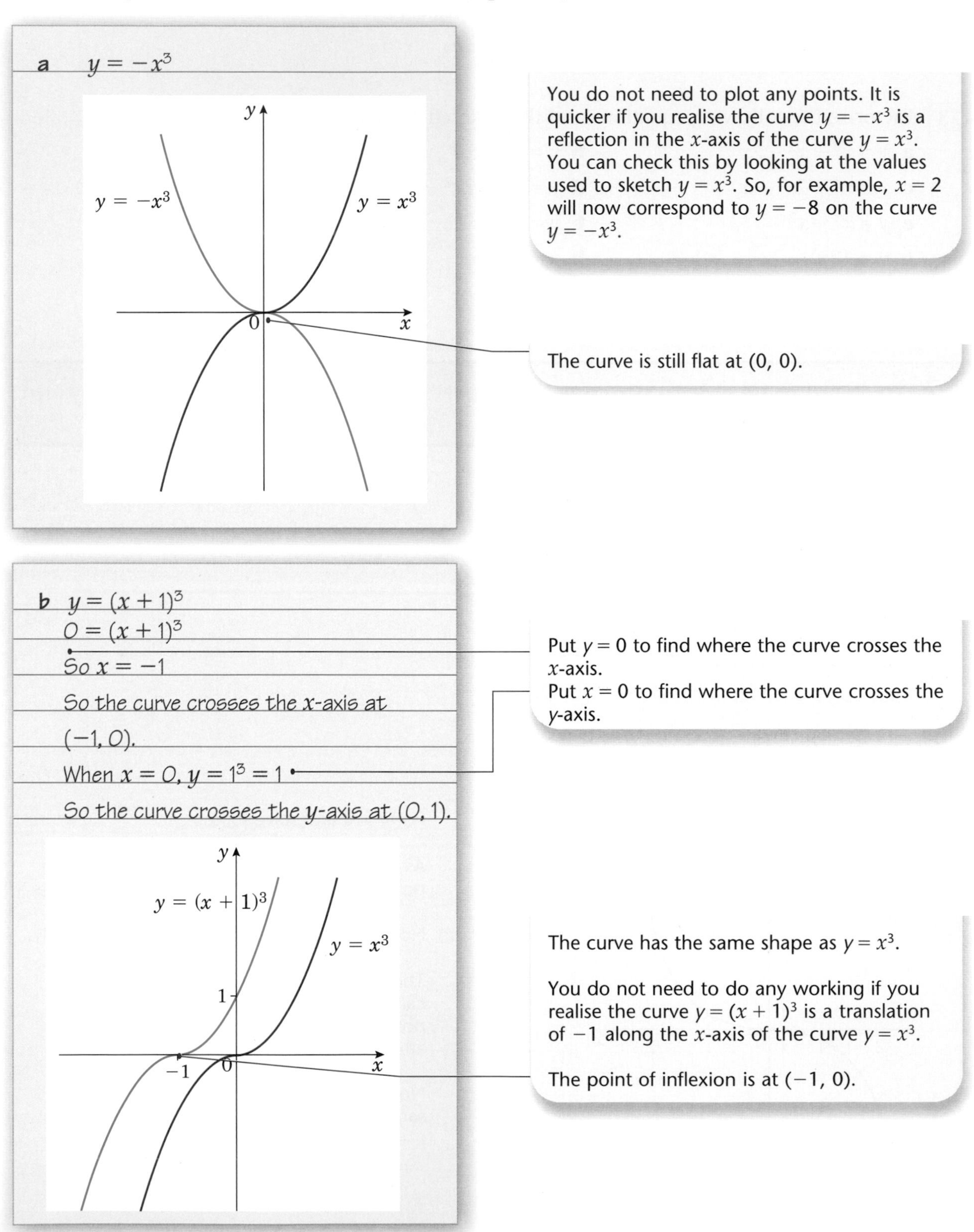

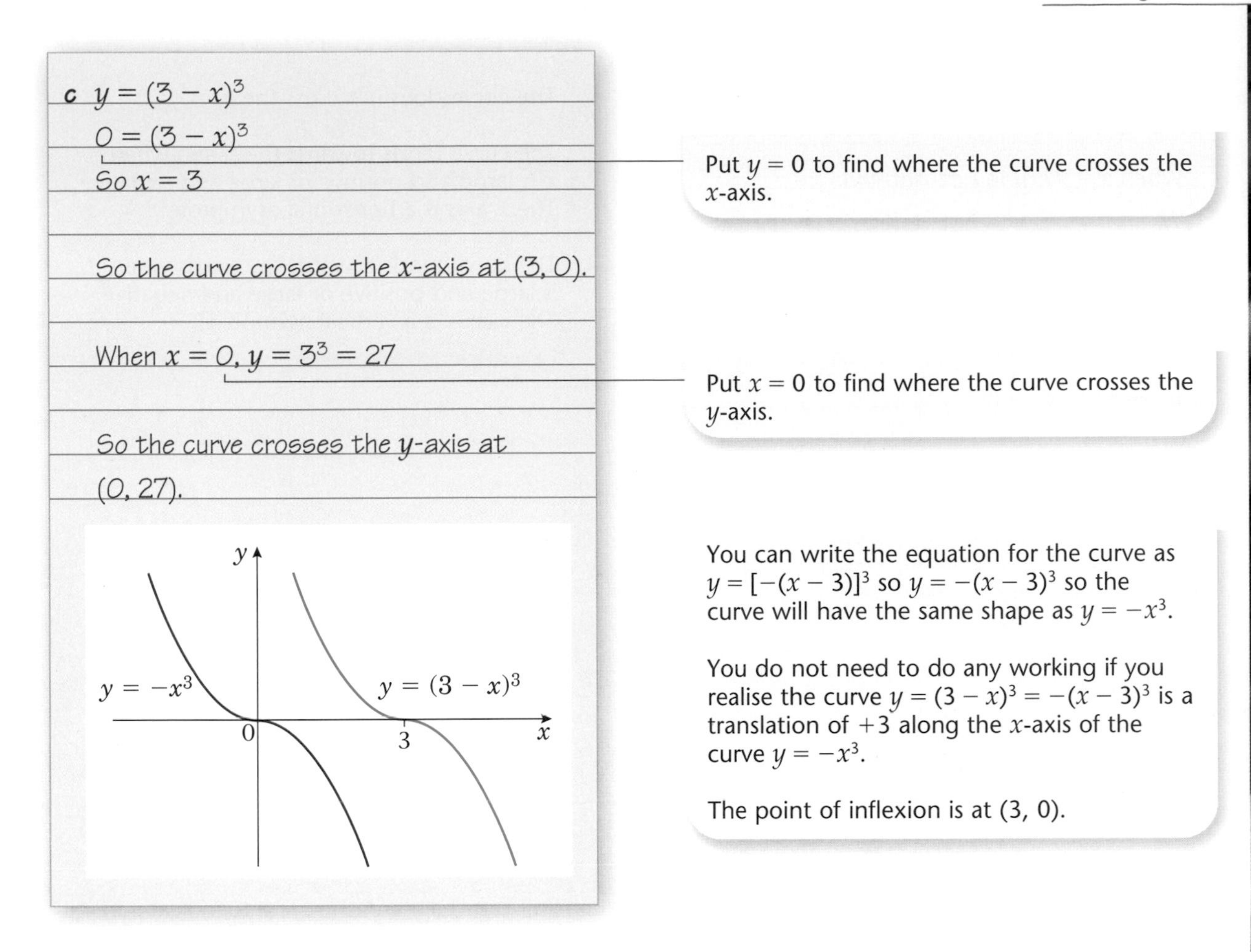

c $y = (3 - x)^3$

$0 = (3 - x)^3$

So $x = 3$

Put $y = 0$ to find where the curve crosses the x-axis.

So the curve crosses the x-axis at $(3, 0)$.

When $x = 0$, $y = 3^3 = 27$

Put $x = 0$ to find where the curve crosses the y-axis.

So the curve crosses the y-axis at $(0, 27)$.

You can write the equation for the curve as $y = [-(x - 3)]^3$ so $y = -(x - 3)^3$ so the curve will have the same shape as $y = -x^3$.

You do not need to do any working if you realise the curve $y = (3 - x)^3 = -(x - 3)^3$ is a translation of $+3$ along the x-axis of the curve $y = -x^3$.

The point of inflexion is at (3, 0).

Exercise 4B

1 Sketch the following curves and show their positions relative to the curve $y = x^3$:

a $y = (x - 2)^3$ **b** $y = (2 - x)^3$ **c** $y = (x - 1)^3$

d $y = (x + 2)^3$ **e** $y = -(x + 2)^3$

2 Sketch the following and indicate the coordinates of the points where the curves cross the axes:

a $y = (x + 3)^3$ **b** $y = (x - 3)^3$ **c** $y = (1 - x)^3$

d $y = -(x - 2)^3$ **e** $y = -(x - \frac{1}{2})^3$

4.3 You need to be able to sketch the reciprocal function $y = \frac{k}{x}$ where k is a constant.

Example 6

Sketch the curve $y = \frac{1}{x}$ and its asymptotes.

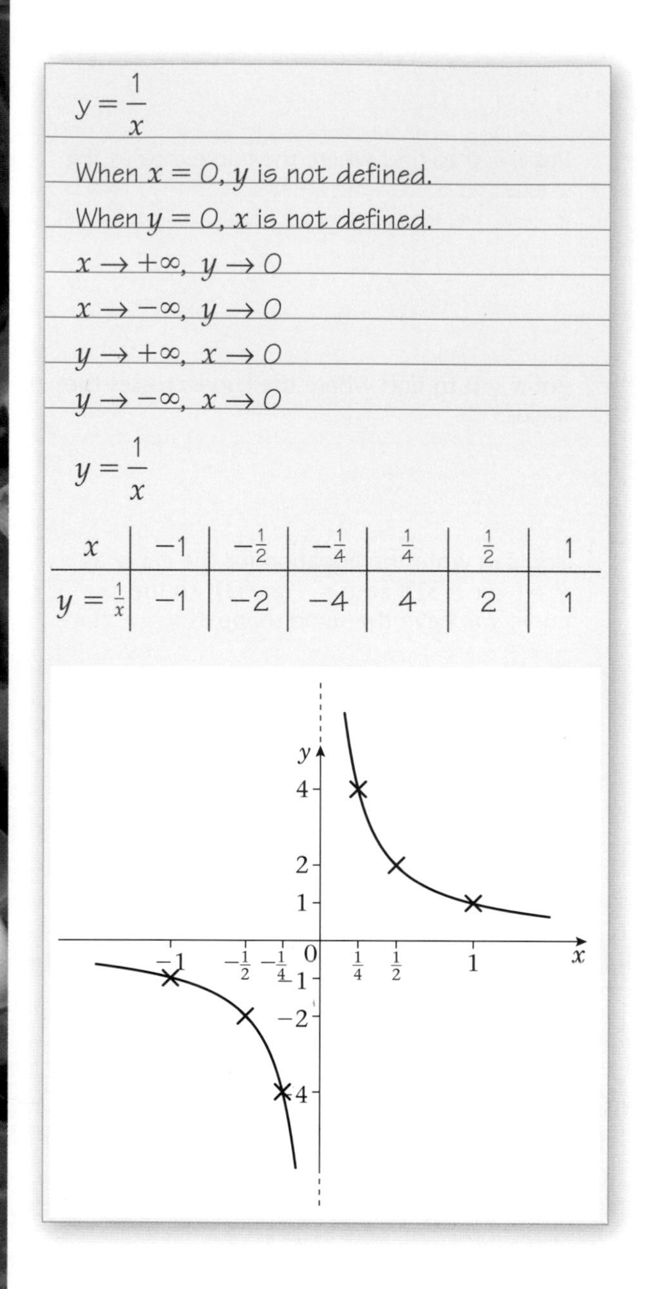

$y = \frac{1}{x}$

When $x = 0$, y is not defined.

When $y = 0$, x is not defined.

$x \to +\infty$, $y \to 0$

$x \to -\infty$, $y \to 0$

$y \to +\infty$, $x \to 0$

$y \to -\infty$, $x \to 0$

$y = \frac{1}{x}$

x	-1	$-\frac{1}{2}$	$-\frac{1}{4}$	$\frac{1}{4}$	$\frac{1}{2}$	1
$y = \frac{1}{x}$	-1	-2	-4	4	2	1

The curve does not cross the axes.

The curve tends towards the x-axis when x is large and positive or large and negative. The x-axis is a horizontal asymptote.

The curve tends towards the y-axis when y is large and positive or large and negative. The y-axis is a vertical asymptote.

The curve does not cross the x-axis or y-axis. You need to plot some points.

You can draw a dashed line to indicate an asymptote. (In this case the asymptotes are the axes, but see Example 11.)

■ **The curves with equations $y = \frac{k}{x}$ fall into two categories:**

Type 1

$y = \frac{k}{x}$, $k > 0$

Type 2

$y = \frac{k}{x}$, $k < 0$

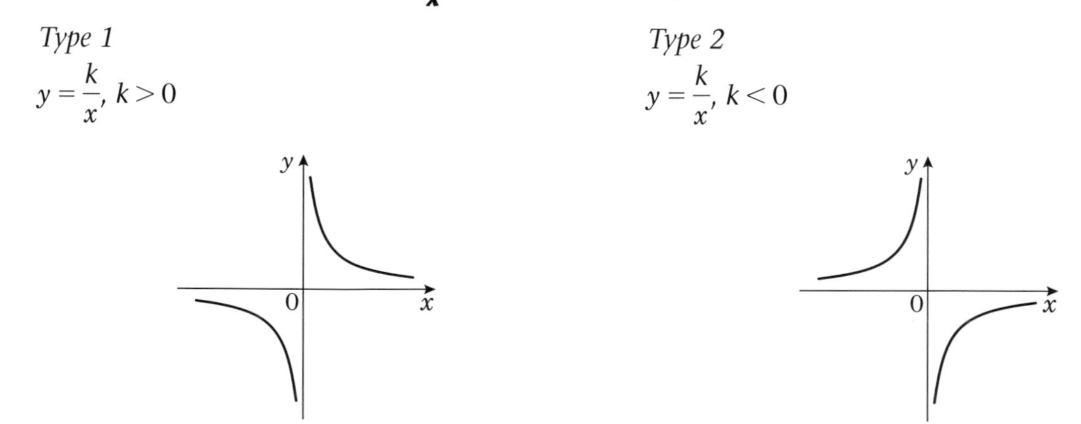

Example 7

Sketch on the same diagram:

a $y = \frac{4}{x}$ and $y = \frac{12}{x}$ **b** $y = -\frac{1}{x}$ and $y = -\frac{3}{x}$

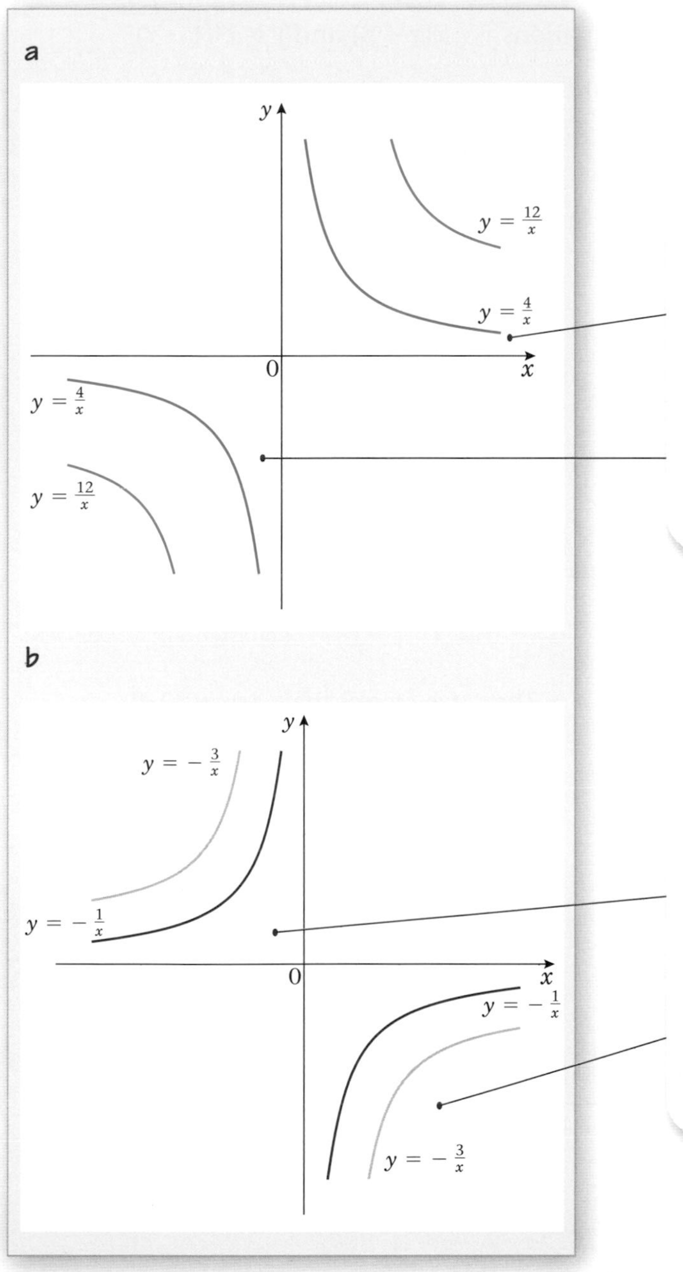

The shape of these curves will be Type 1.

In this quadrant, $x > 0$ so for any values of x: $\frac{12}{x} > \frac{4}{x}$

In this quadrant, $x < 0$ so for any values of x: $\frac{12}{x} < \frac{4}{x}$

The shape of these curves will be Type 2.

In this quadrant, $x < 0$ so for any values of x: $\frac{-3}{x} > \frac{-1}{x}$

In this quadrant, $x > 0$ so for any values of x: $\frac{-3}{x} < \frac{-1}{x}$

Exercise 4C

Use a separate diagram to sketch each pair of graphs.

1 $y = \frac{2}{x}$ and $y = \frac{4}{x}$

2 $y = \frac{2}{x}$ and $y = -\frac{2}{x}$

3 $y = -\frac{4}{x}$ and $y = -\frac{2}{x}$

4 $y = \frac{3}{x}$ and $y = \frac{8}{x}$

5 $y = -\frac{3}{x}$ and $y = -\frac{8}{x}$

4.4 You can sketch curves of functions to show points of intersection and solutions to equations.

Example 8

a On the same diagram sketch the curves with equations $y = x(x - 3)$ and $y = x^2(1 - x)$.

b Find the coordinates of the point of intersection.

a $y = x(x - 3)$

$0 = x(x - 3)$

Put $y = 0$ and solve for x.

So $x = 0$ or $x = 3$.

So the curve crosses the x-axis at $(0, 0)$ and $(3, 0)$.

$y = x^2(1 - x)$

$0 = x^2(1 - x)$

Put $y = 0$ and solve for x to find where the curve crosses the x-axis.

So $x = 0$ or $x = 1$.

So the curve crosses the x-axis at $(0, 0)$ or $(1, 0)$.

The curve crosses the y-axis at $(0, 0)$.

$x \to \infty$, $y \to -\infty$

$x \to -\infty$, $y \to +\infty$

Check what happens to y for large positive and negative values of x.

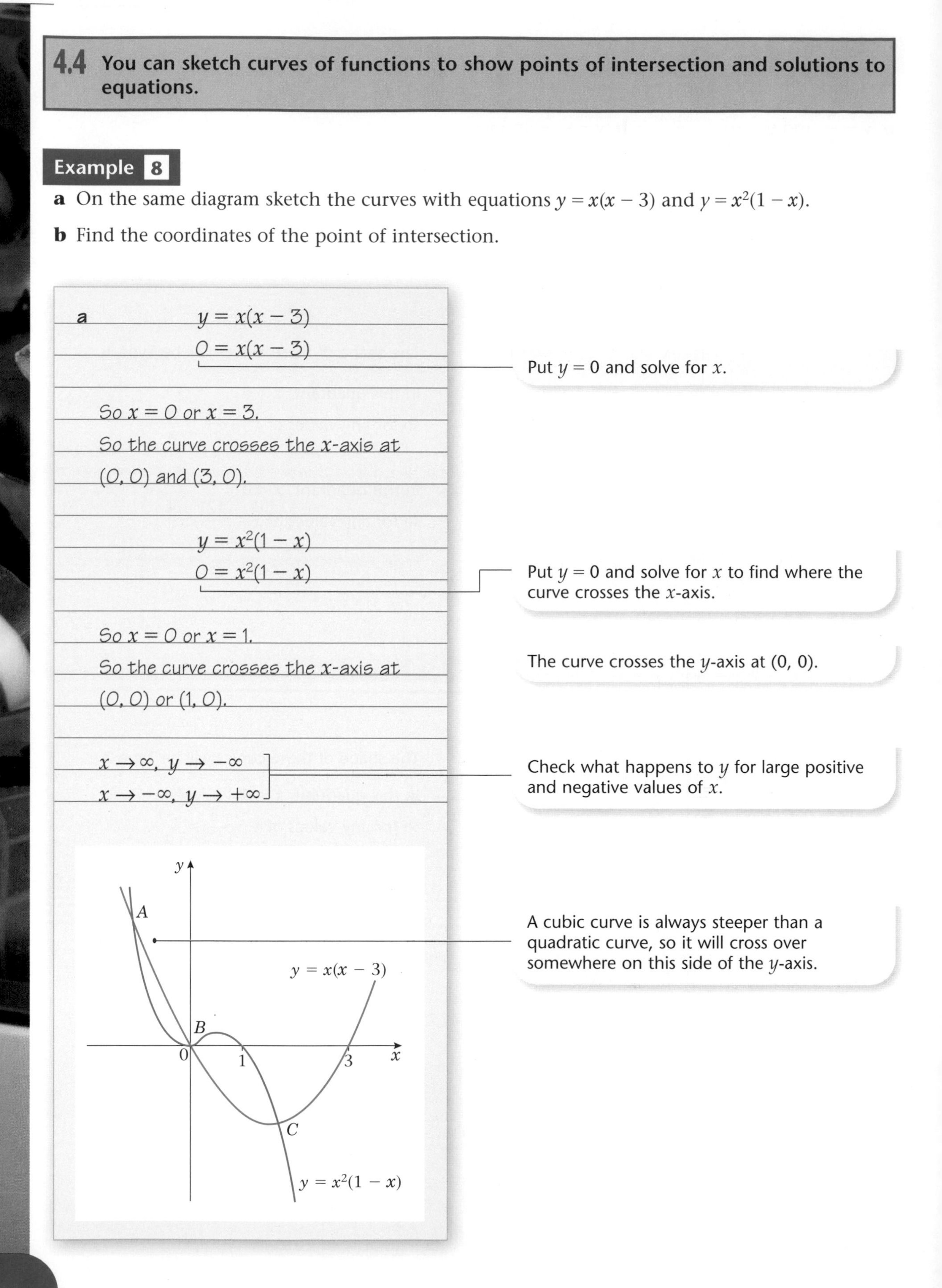

A cubic curve is always steeper than a quadratic curve, so it will cross over somewhere on this side of the y-axis.

b From the graph there are three points where the curves cross, labelled A, B and C. The x-coordinates are given by the solutions to the equation.

$$x(x-3) = x^2(1-x)$$

$$x^2 - 3x = x^2 - x^3$$

Multiply out brackets (see Section 1.3).

$$x^3 - 3x = 0$$

Collect terms on one side.

$$x(x^2 - 3) = 0$$

Factorise.

$$x(x-\sqrt{3})(x+\sqrt{3}) = 0$$

Factorise using a difference of 2 squares.

So $x = -\sqrt{3}, 0, \sqrt{3}$

You can use the equation $y = x^2(1-x)$ to find the y-coordinates.

So the point where x is negative is $A\,(-\sqrt{3}, 3[1+\sqrt{3}])$, B is $(0, 0)$ and C is the point $(\sqrt{3}, 3[1-\sqrt{3}])$.

Example 9

a On the same diagram sketch the curves with equations $y = x^2(x-1)$ and $y = \dfrac{2}{x}$.

b Explain how your sketch shows that there are two solutions to the equation $x^2(x-1) - \dfrac{2}{x} = 0$.

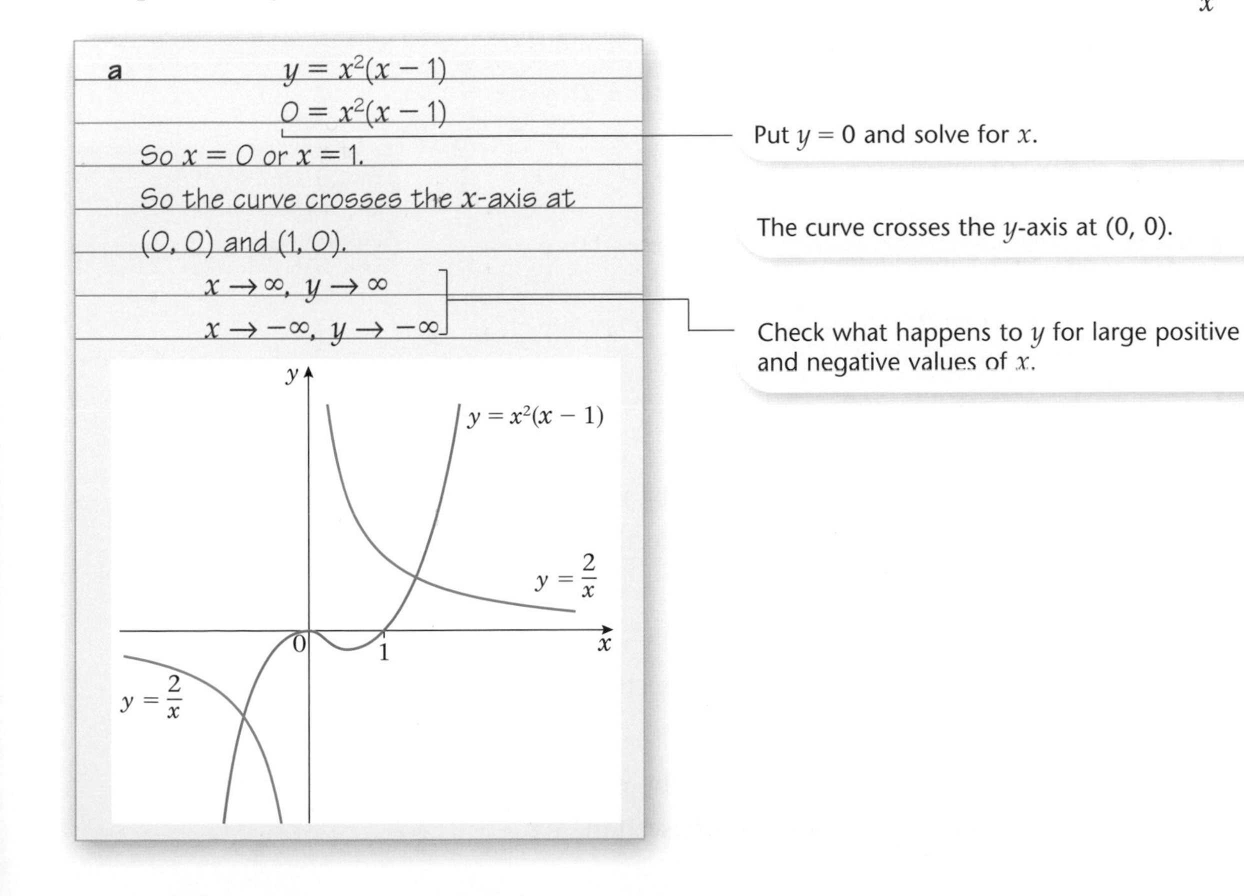

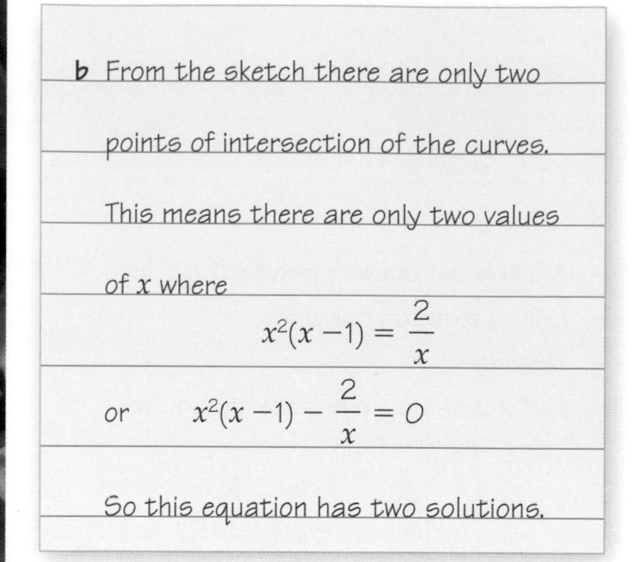

b From the sketch there are only two points of intersection of the curves.

This means there are only two values of x where

$$x^2(x-1) = \frac{2}{x}$$

or $$x^2(x-1) - \frac{2}{x} = 0$$

So this equation has two solutions.

You would not be expected to solve this equation in C1.

Exercise 4D

1 In each case:

i sketch the two curves on the same axes

ii state the number of points of intersection

iii write down a suitable equation which would give the x-coordinates of these points. (You are not required to solve this equation.)

a $y = x^2$, $y = x(x^2 - 1)$

b $y = x(x + 2)$, $y = -\dfrac{3}{x}$

c $y = x^2$, $y = (x + 1)(x - 1)^2$

d $y = x^2(1 - x)$, $y = -\dfrac{2}{x}$

e $y = x(x - 4)$, $y = \dfrac{1}{x}$

f $y = x(x - 4)$, $y = -\dfrac{1}{x}$

g $y = x(x - 4)$, $y = (x - 2)^3$

h $y = -x^3$, $y = -\dfrac{2}{x}$

i $y = -x^3$, $y = x^2$

j $y = -x^3$, $y = -x(x + 2)$

Hint: In question 1f, check the point $x = 2$ in both curves.

2 **a** On the same axes sketch the curves given by $y = x^2(x - 4)$ and $y = x(4 - x)$.

b Find the coordinates of the points of intersection.

3 **a** On the same axes sketch the curves given by $y = x(2x + 5)$ and $y = x(1 + x)^2$

b Find the coordinates of the points of intersection.

4 **a** On the same axes sketch the curves given by $y = (x - 1)^3$ and $y = (x - 1)(1 + x)$.

b Find the coordinates of the points of intersection.

5 **a** On the same axes sketch the curves given by $y = x^2$ and $y = -\frac{27}{x}$.

b Find the coordinates of the point of intersection.

6 **a** On the same axes sketch the curves given by $y = x^2 - 2x$ and $y = x(x - 2)(x - 3)$.

b Find the coordinates of the point of intersection.

7 **a** On the same axes sketch the curves given by $y = x^2(x - 3)$ and $y = \frac{2}{x}$.

b Explain how your sketch shows that there are only two solutions to the equation $x^3(x - 3) = 2$.

8 **a** On the same axes sketch the curves given by $y = (x + 1)^3$ and $y = 3x(x - 1)$.

b Explain how your sketch shows that there is only one solution to the equation $x^3 + 6x + 1 = 0$.

9 **a** On the same axes sketch the curves given by $y = \frac{1}{x}$ and $y = -x(x - 1)^2$.

b Explain how your sketch shows that there are no solutions to the equation $1 + x^2(x - 1)^2 = 0$.

10 **a** On the same axes sketch the curves given by $y = 1 - 4x^2$ and $y = x(x - 2)^2$.

b State, with a reason, the number of solutions to the equation $x^3 + 4x - 1 = 0$.

11 **a** On the same axes sketch the curve $y = x^3 - 3x^2 - 4x$ and the line $y = 6x$.

b Find the coordinates of the points of intersection.

12 **a** On the same axes sketch the curve $y = (x^2 - 1)(x - 2)$ and the line $y = 14x + 2$.

b Find the coordinates of the points of intersection.

13 **a** On the same axes sketch the curves with equations $y = (x - 2)(x + 2)^2$ and $y = -x^2 - 8$.

b Find the coordinates of the points of intersection.

4.5 **You can transform the curve of a function f(x) by simple translations of the form:**

- **f($x + a$) is a horizontal translation of $-a$**
- **f(x) + a is a vertical translation of $+a$.**

Example 10

Sketch the curves for:

a $f(x) = x^2$ **b** $g(x) = (x - 2)^2$ **c** $h(x) = x^2 + 2$

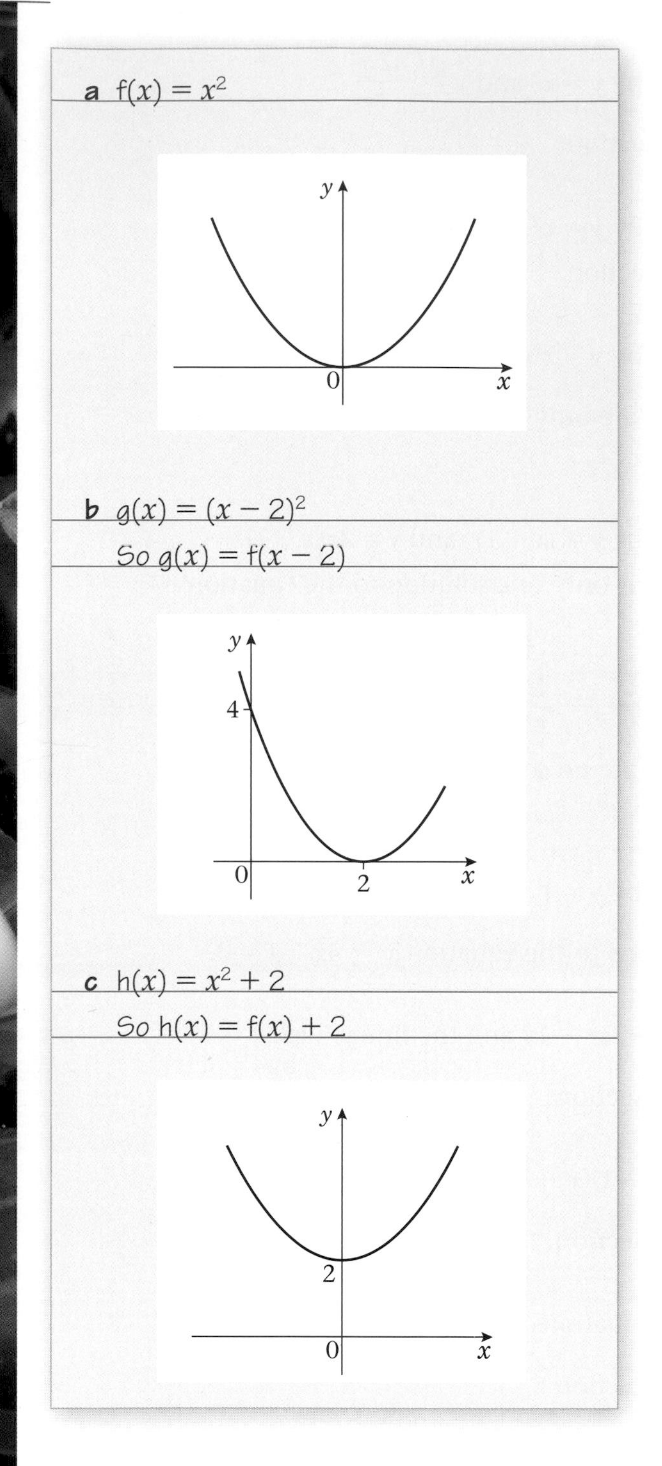

Here $a = -2$ so g(x) is a horizontal translation of $-(-2) = +2$ along the x-axis.

Here $a = +2$ so h(x) is a vertical translation of $+2$ along the y-axis.

Example 11

a Given that **i** $f(x) = x^3$

ii $g(x) = x(x - 2)$,

sketch the curves with equation $y = f(x + 1)$ and $g(x + 1)$ and mark on your sketch the points where the curves cross the axes.

b Given that $h(x) = \dfrac{1}{x}$, sketch the curve with equation $y = h(x) + 1$ and state the equations of any asymptotes and intersections with the axes.

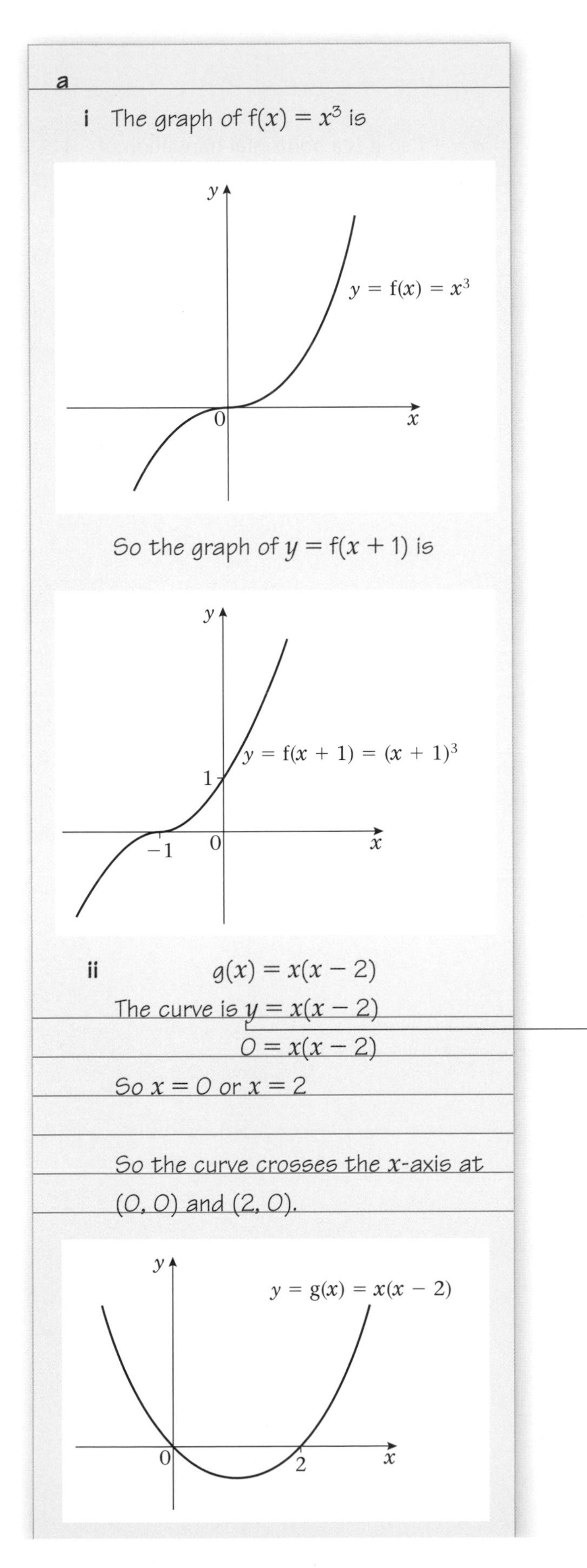

First sketch $f(x)$.

Here $a = +1$ so it is a horizontal translation of -1 along the x-axis.

In this case the new equations can easily be found as $y = (x + 1)^3$ and this may help with the sketch.

Put $y = 0$ to find where the curve crosses the x-axis.

First sketch $g(x)$.

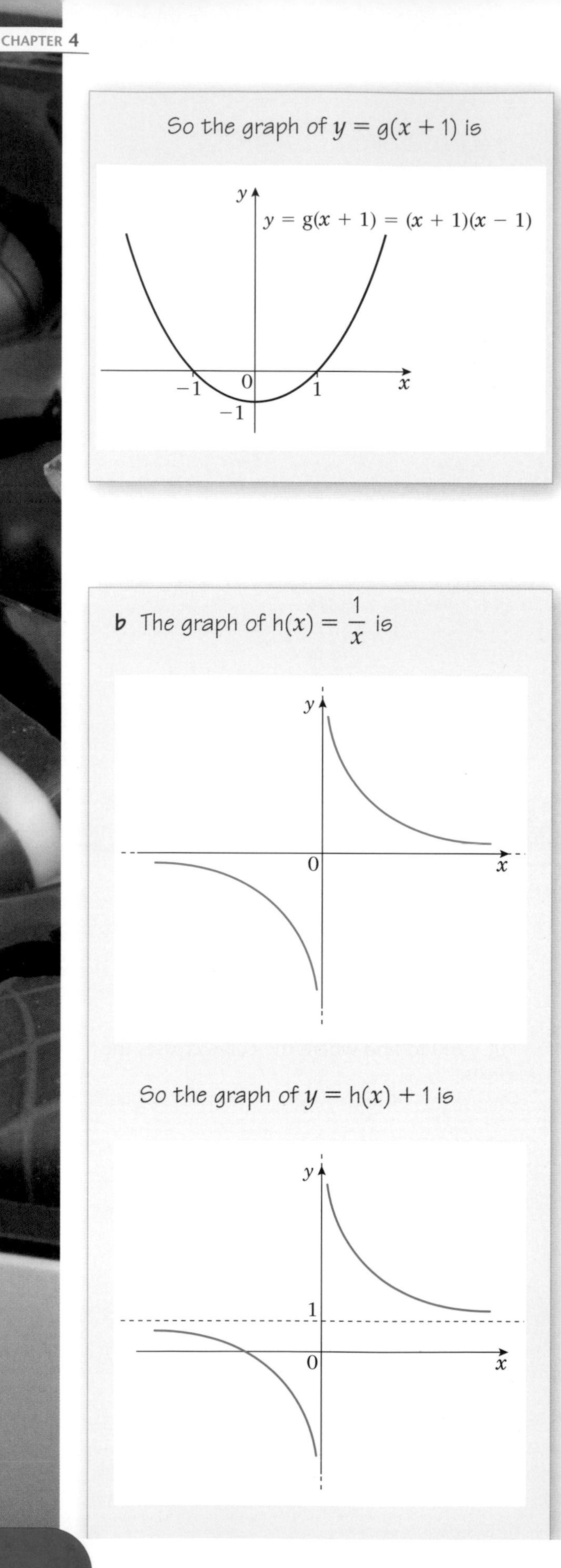

$a = +1$ so it is a horizontal translation of -1 along the x-axis.

You find the equation for $g(x + 1)$ by replacing x by $(x + 1)$ in the original equation. So $y = g(x + 1) = (x + 1)(x + 1 - 2) = (x + 1)(x - 1)$.

You can see this matches your sketch. The intersection with the y-axis is now at $(0, -1)$.

First sketch $h(x)$.

Here $a = +1$ so it is a vertical translation of $+1$ along the y-axis.

The curve crosses the x-axis once.

$y = \mathrm{h}(x) + 1 = \frac{1}{x} + 1$

$0 = \frac{1}{x} + 1$

$-1 = \frac{1}{x}$

$x = -1$

So the curve intersects the x-axis at $(-1, 0)$.

The horizontal asymptote is $y = 1$.

The vertical asymptote is $x = 0$.

Put $y = 0$ to find where the curve crosses the x-axis.

Exercise 4E

1 Apply the following transformations to the curves with equations $y = \mathrm{f}(x)$ where:

i $\mathrm{f}(x) = x^2$ **ii** $\mathrm{f}(x) = x^3$ **iii** $\mathrm{f}(x) = \frac{1}{x}$

In each case state the coordinates of points where the curves cross the axes and in **iii** state the equations of any asymptotes.

a $\mathrm{f}(x + 2)$ **b** $\mathrm{f}(x) + 2$ **c** $\mathrm{f}(x - 1)$

d $\mathrm{f}(x) - 1$ **e** $\mathrm{f}(x) - 3$ **f** $\mathrm{f}(x - 3)$

2 **a** Sketch the curve $y = \mathrm{f}(x)$ where $\mathrm{f}(x) = (x - 1)(x + 2)$.

b On separate diagrams sketch the graphs of **i** $y = \mathrm{f}(x + 2)$ **ii** $y = \mathrm{f}(x) + 2$.

c Find the equations of the curves $y = \mathrm{f}(x + 2)$ and $y = \mathrm{f}(x) + 2$, in terms of x, and use these equations to find the coordinates of the points where your graphs in part **b** cross the y-axis.

3 **a** Sketch the graph of $y = \mathrm{f}(x)$ where $\mathrm{f}(x) = x^2(1 - x)$.

b Sketch the curve with equation $y = \mathrm{f}(x + 1)$.

c By finding the equation $\mathrm{f}(x + 1)$ in terms of x, find the coordinates of the point in part **b** where the curve crosses the y-axis.

4 **a** Sketch the graph of $y = \mathrm{f}(x)$ where $\mathrm{f}(x) = x(x - 2)^2$.

b Sketch the curves with equations $y = \mathrm{f}(x) + 2$ and $y = \mathrm{f}(x + 2)$.

c Find the coordinates of the points where the graph of $y = \mathrm{f}(x + 2)$ crosses the axes.

5 **a** Sketch the graph of $y = \mathrm{f}(x)$ where $\mathrm{f}(x) = x(x - 4)$.

b Sketch the curves with equations $y = \mathrm{f}(x + 2)$ and $y = \mathrm{f}(x) + 4$.

c Find the equations of the curves in part **b** in terms of x and hence find the coordinates of the points where the curves cross the axes.

4.6 **You can transform the curve of a function f(x) by simple stretches of these forms:**

- **f(ax) is a horizontal stretch of scale factor $\frac{1}{a}$, so you multiply the x-coordinates by $\frac{1}{a}$ and leave the y-coordinates unchanged.**
- **af(x) is a vertical stretch of scale factor a, so you multiply the y-coordinates by a and leave the x-coordinates unchanged.**

Example 12

Given that $f(x) = 9 - x^2$, sketch the curves with equations:

a $y = f(2x)$ **b** $y = 2f(x)$

a $f(x) = 9 - x^2$

So $f(x) = (3 - x)(3 + x)$

You can factorise the expression.

The curve is $y = (3 - x)(3 + x)$

$0 = (3 - x)(3 + x)$

So $x = 3$ or $x = -3$

Put $y = 0$ to find where the curve crosses the x-axis.

So the curve crosses the x-axis at $(3, 0)$ and $(-3, 0)$.

When $x = 0$, $y = 3 \times 3 = 9$

So the curve crosses the y-axis at $(0, 9)$.

Put $x = 0$ to find where the curve crosses the y-axis.

The curve $y = f(x)$ is

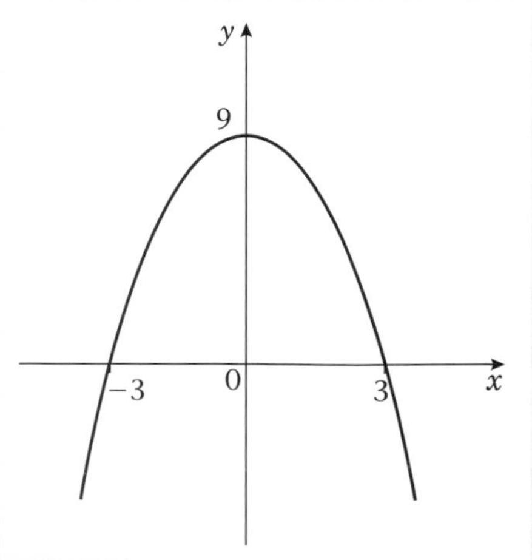

First sketch $y = f(x)$.

$y = f(2x)$ so the curve is

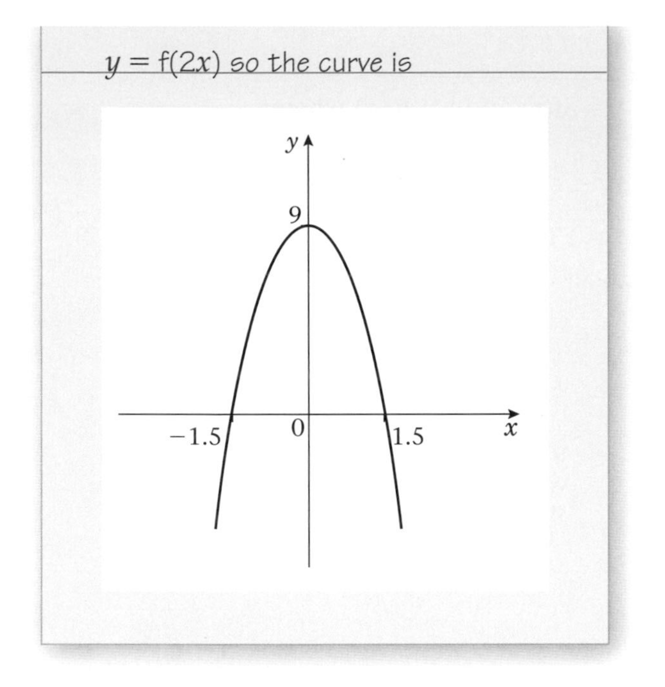

$y = f(ax)$ where $a = 2$ so it is a horizontal stretch with scale factor $\frac{1}{2}$.

Check: The curve is $y = f(2x)$.
So $y = (3 - 2x)(3 + 2x)$.
When $y = 0$, $x = -1.5$ or $x = 1.5$.
So the curve crosses the x-axis at $(-1.5, 0)$ and $(1.5, 0)$.
When $x = 0$, $y = 9$.
So the curve crosses the y-axis at $(0, 9)$.

b $y = 2f(x)$
So the curve is

$y = af(x)$ where $a = 2$ so it is a vertical stretch with scale factor 2.

Check: The curve is $y = 2f(x)$.
So $y = 2(3 - x)(3 + x)$.
When $y = 0$, $x = 3$ or $x = -3$.
So the curve crosses the x-axis at $(-3, 0)$ and $(3, 0)$.
When $x = 0$, $y = 2 \times 9 = 18$.
So the curve crosses the y-axis at $(0, 18)$.

Example 13

a On the same axes sketch the graphs of $y = f(x)$, $y = 3f(x)$ and $y = f(\frac{1}{3}x)$ where:

i $f(x) = x^3$ **ii** $f(x) = \dfrac{1}{x}$

b On the same axes sketch the graphs of $y = f(x)$, $y = -f(x)$ and $y = f(-x)$ where $f(x) = x(x + 2)$.

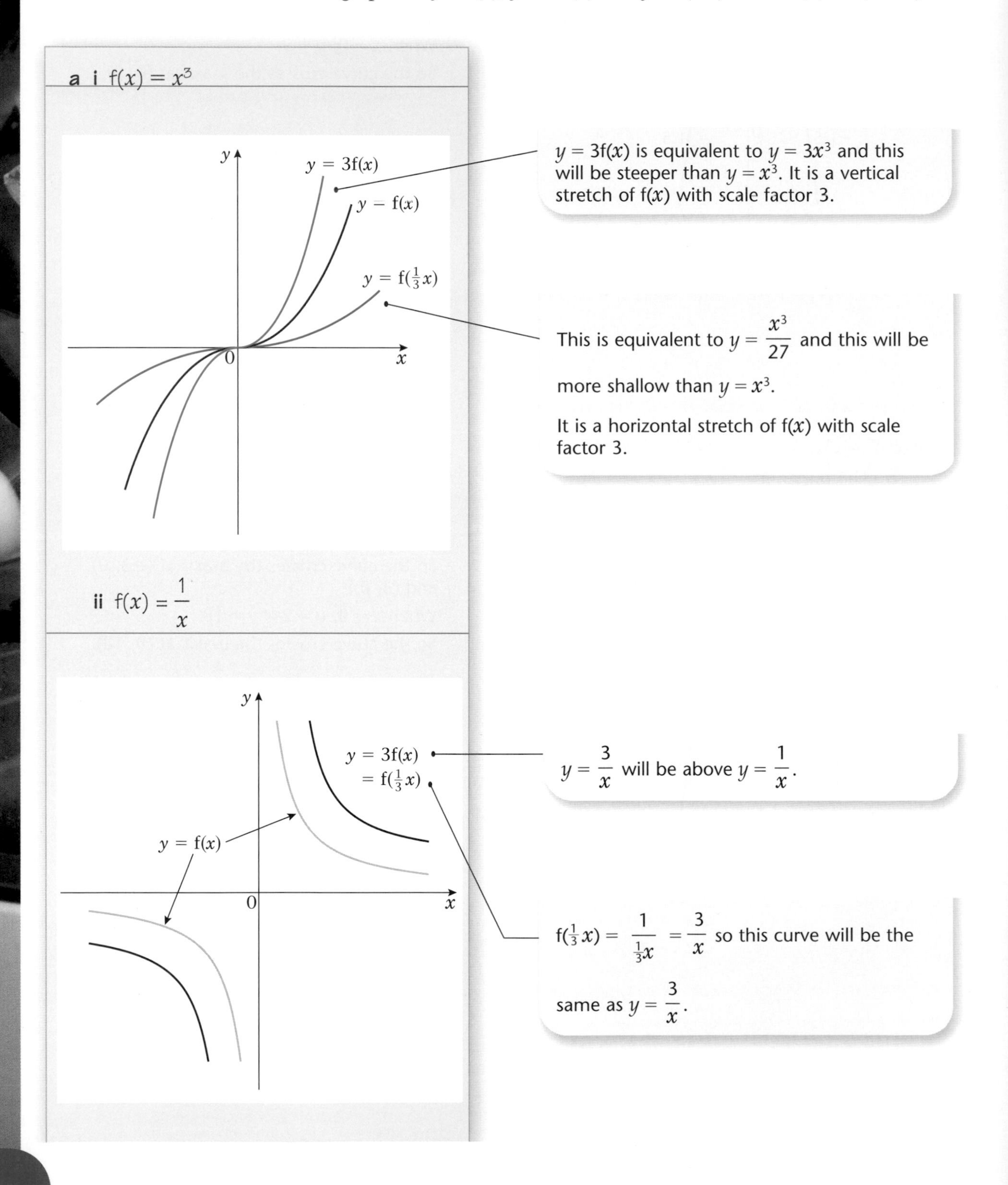

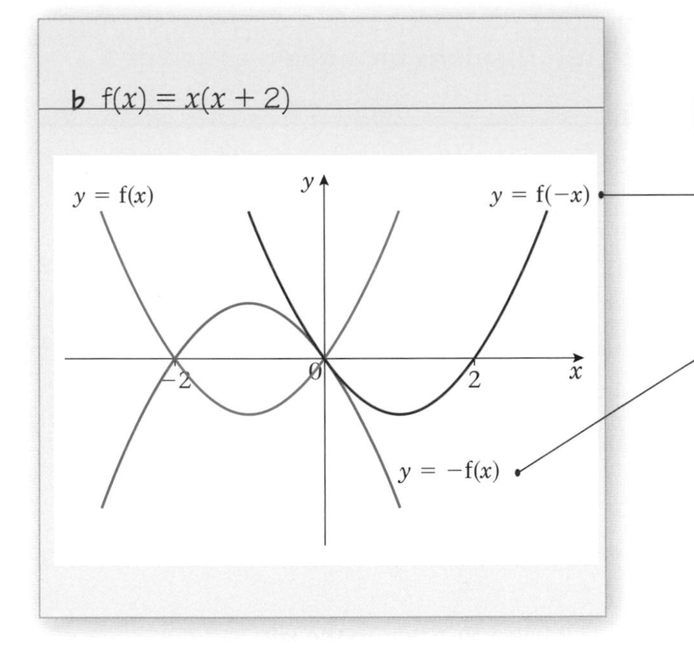

$y = f(-x)$ is $y = (-x)(-x + 2)$ which is $y = x^2 - 2x$ or $y = x(x - 2)$ and this is a reflection of the original curve in the y-axis. Alternatively multiply each x-coordinate by -1 and leave the y coordinates unchanged.

$y = -f(x)$ is $y = -x(x + 2)$ and this is a reflection of the original curve in the x-axis. Alternatively simply remember each y-coordinate is multiplied by -1 and the x-coordinates remain unchanged.

Exercise 4F

1 Apply the following transformations to the curves with equations $y = f(x)$ where:

i $f(x) = x^2$ **ii** $f(x) = x^3$ **iii** $f(x) = \dfrac{1}{x}$

In each case show both $f(x)$ and the transformation on the same diagram.

a $f(2x)$ **b** $f(-x)$

c $f(\frac{1}{2}x)$ **d** $f(4x)$

e $f(\frac{1}{4}x)$ **f** $2f(x)$

g $-f(x)$ **h** $4f(x)$

i $\frac{1}{2}f(x)$ **j** $\frac{1}{4}f(x)$

2 **a** Sketch the curve with equation $y = f(x)$ where $f(x) = x^2 - 4$.

b Sketch the graphs of $y = f(4x)$, $y = 3f(x)$, $y = f(-x)$ and $y = -f(x)$.

3 **a** Sketch the curve with equation $y = f(x)$ where $f(x) = (x - 2)(x + 2)x$.

b Sketch the graphs of $y = f(\frac{1}{2}x)$, $y = f(2x)$ and $y = -f(x)$.

4 **a** Sketch the curve with equation $y = f(x)$ where $f(x) = x^2(x - 3)$.

b Sketch the curves with equations $y = f(2x)$, $y = -f(x)$ and $y = f(-x)$.

5 **a** Sketch the curve with equation $y = f(x)$ where $f(x) = (x - 2)(x - 1)(x + 2)$.

b Sketch the curves with equations $y = f(2x)$ and $f(\frac{1}{2}x)$.

4.7 You need to be able to perform simple transformations on a given sketch of a function.

Example 14

The following diagram shows a sketch of the curve f(x) which passes through the origin. The points $A(1, 4)$ and $B(3, 1)$ also lie on the curve.

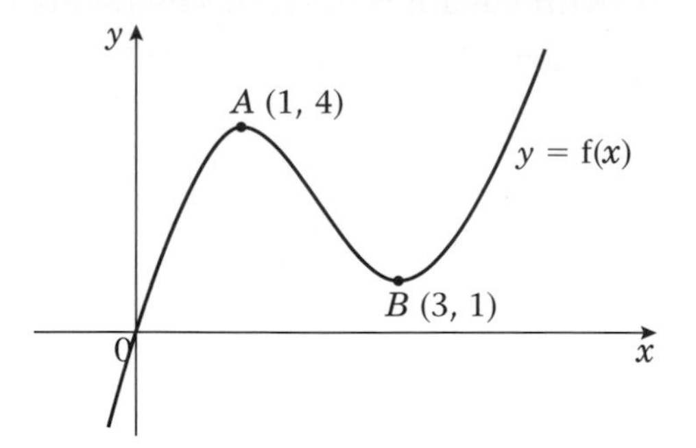

Sketch the following:

a $y = \text{f}(x + 1)$ **b** $y = \text{f}(x - 1)$ **c** $y = \text{f}(x) - 4$

In each case you should show the coordinates of the images of the points O, A and B.

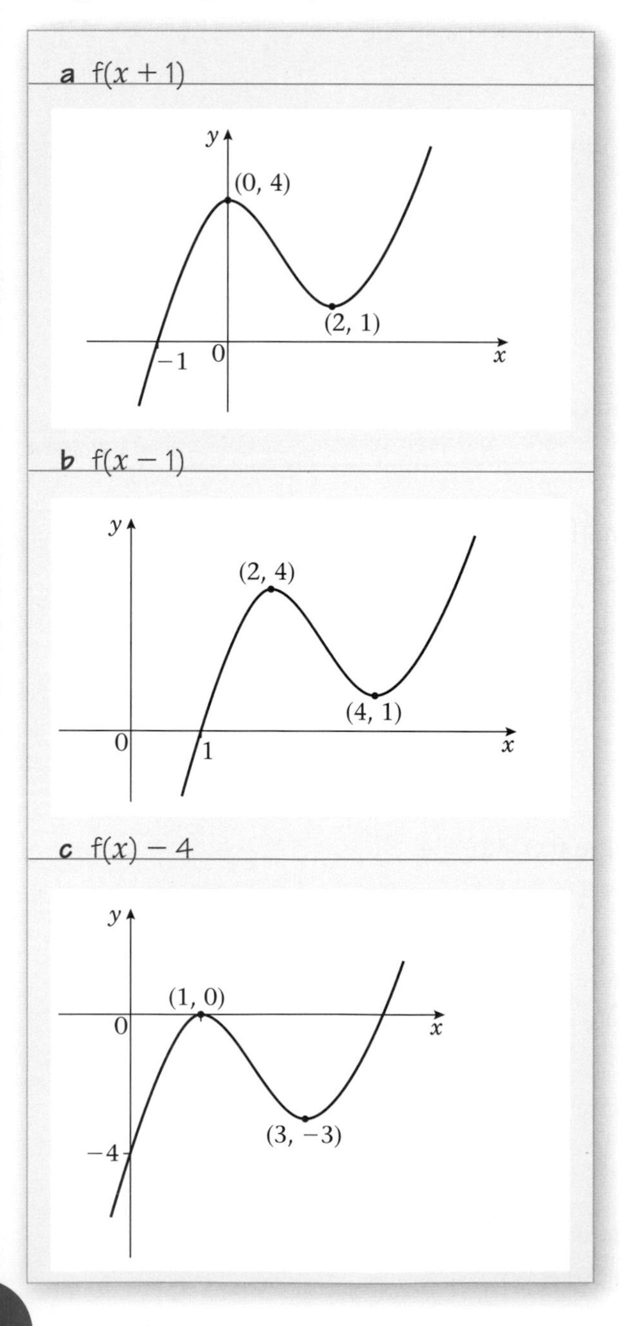

Move f(x) 1 unit to the left.

This means move f(x) 1 unit to the right.

Move f(x) down 4 units.

Exercise 4G

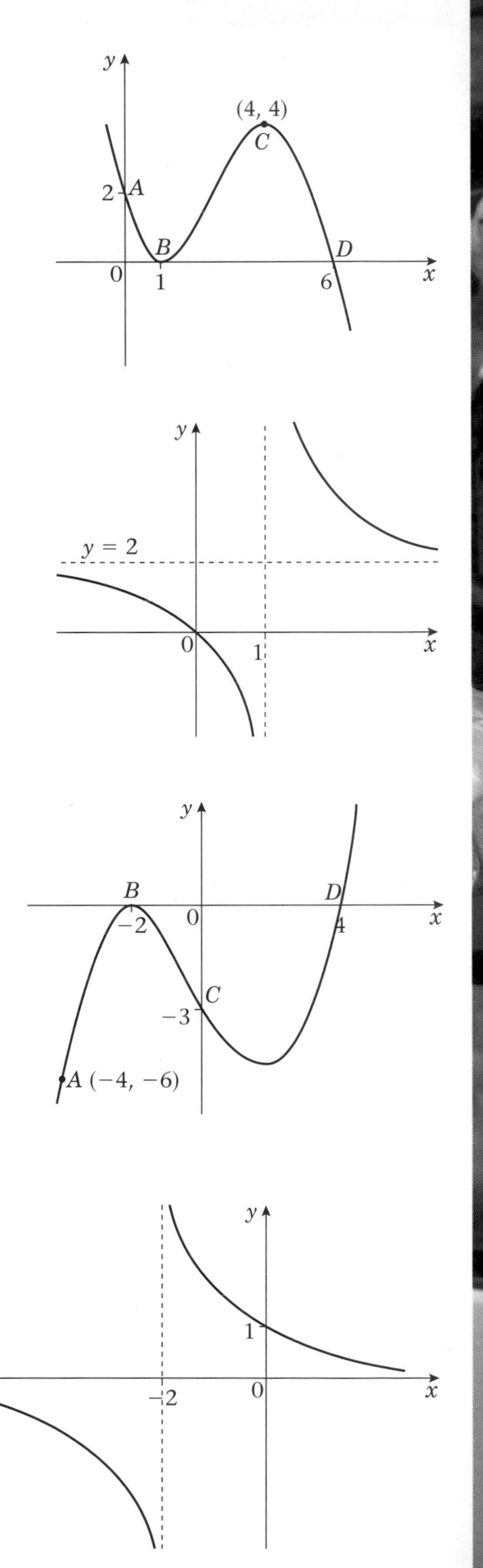

1 The following diagram shows a sketch of the curve with equation $y = f(x)$. The points $A(0, 2)$, $B(1, 0)$, $C(4, 4)$ and $D(6, 0)$ lie on the curve.

Sketch the following graphs and give the coordinates of the points A, B, C and D after each transformation:

a $f(x + 1)$ **b** $f(x) - 4$ **c** $f(x + 4)$
d $f(2x)$ **e** $3f(x)$ **f** $f(\frac{1}{2}x)$
g $\frac{1}{2}f(x)$ **h** $f(-x)$

2 The curve $y = f(x)$ passes through the origin and has horizontal asymptote $y = 2$ and vertical asymptote $x = 1$, as shown in the diagram.

Sketch the following graphs and give the equations of any asymptotes and, for all graphs except **a**, give coordinates of intersections with the axes after each transformation.

a $f(x) + 2$ **b** $f(x + 1)$ **c** $2f(x)$
d $f(x) - 2$ **e** $f(2x)$ **f** $f(\frac{1}{2}x)$
g $\frac{1}{2}f(x)$ **h** $-f(x)$

3 The curve with equation $y = f(x)$ passes through the points $A(-4, -6)$, $B(-2, 0)$, $C(0, -3)$ and $D(4, 0)$ as shown in the diagram.

Sketch the following and give the coordinates of the points A, B, C and D after each transformation.

a $f(x - 2)$ **b** $f(x) + 6$ **c** $f(2x)$
d $f(x + 4)$ **e** $f(x) + 3$ **f** $3f(x)$
g $\frac{1}{3}f(x)$ **h** $f(\frac{1}{4}x)$ **i** $-f(x)$
j $f(-x)$

4 A sketch of the curve $y = f(x)$ is shown in the diagram. The curve has a vertical asymptote with equation $x = -2$ and a horizontal asymptote with equation $y = 0$. The curve crosses the y-axis at $(0, 1)$.

a Sketch, on separate diagrams, the graphs of:
i $2f(x)$ **ii** $f(2x)$ **iii** $f(x - 2)$
iv $f(x) - 1$ **v** $f(-x)$ **vi** $-f(x)$
In each case state the equations of any asymptotes and, if possible, points where the curve cuts the axes.

b Suggest a possible equation for $f(x)$.

Mixed exercise 4H

1 **a** On the same axes sketch the graphs of $y = x^2(x - 2)$ and $y = 2x - x^2$.

b By solving a suitable equation find the points of intersection of the two graphs.

2 **a** On the same axes sketch the curves with equations $y = \frac{6}{x}$ and $y = 1 + x$.

b The curves intersect at the points A and B. Find the coordinates of A and B.

c The curve C with equation $y = x^2 + px + q$, where p and q are integers, passes through A and B. Find the values of p and q.

d Add C to your sketch.

3 The diagram shows a sketch of the curve $y = \text{f}(x)$. The point $B(0, 0)$ lies on the curve and the point $A(3, 4)$ is a maximum point. The line $y = 2$ is an asymptote.

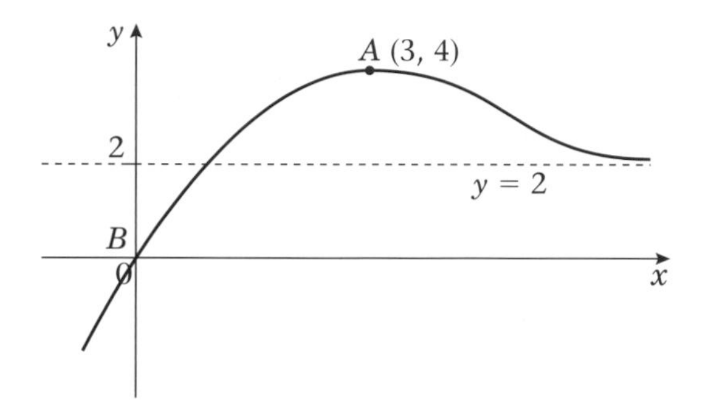

Sketch the following and in each case give the coordinates of the new positions of A and B and state the equation of the asymptote:

a $\text{f}(2x)$ **b** $\frac{1}{2}\text{f}(x)$ **c** $\text{f}(x) - 2$

d $\text{f}(x + 3)$ **e** $\text{f}(x - 3)$ **f** $\text{f}(x) + 1$

4 The diagram shows the curve with equation $y = 5 + 2x - x^2$ and the line with equation $y = 2$. The curve and the line intersect at the points A and B.

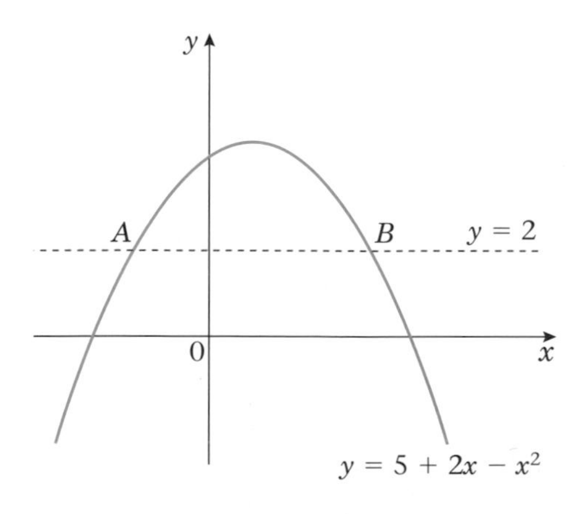

Find the x-coordinates of A and B.

E

5 The curve with equation $y = \mathrm{f}(x)$ meets the coordinate axes at the points $(-1, 0)$, $(4, 0)$ and $(0, 3)$, as shown in the diagram.

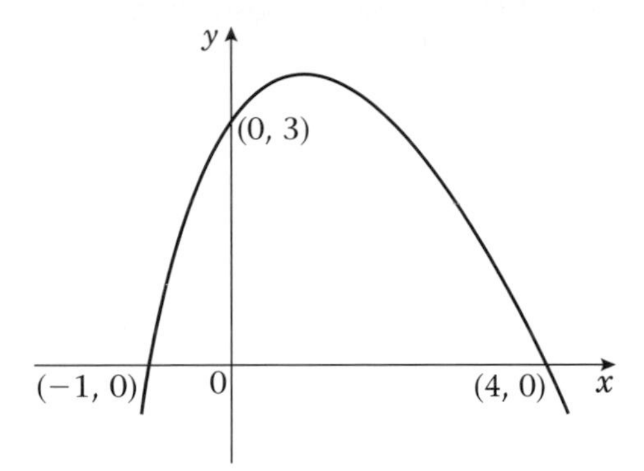

Using a separate diagram for each, sketch the curve with equation

a $y = \mathrm{f}(x - 1)$ **b** $y = -\mathrm{f}(x)$

On each sketch, write in the coordinates of the points at which the curve meets the coordinate axes.

E

6 The figure shows a sketch of the curve with equation $y = \mathrm{f}(x)$.

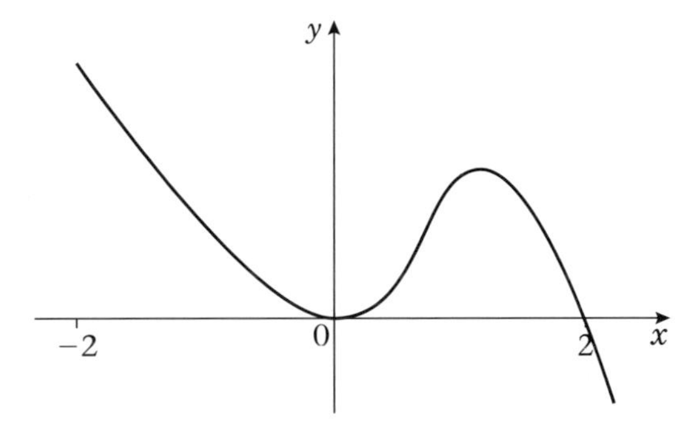

In separate diagrams show, for $-2 \leqslant x \leqslant 2$, sketches of the curves with equation:

a $y = \mathrm{f}(-x)$ **b** $y = -\mathrm{f}(x)$

Mark on each sketch the x-coordinate of any point, or points, where a curve touches or crosses the x-axis.

E

7 The diagram shows the graph of the quadratic function f. The graph meets the x-axis at $(1, 0)$ and $(3, 0)$ and the minimum point is $(2, -1)$.

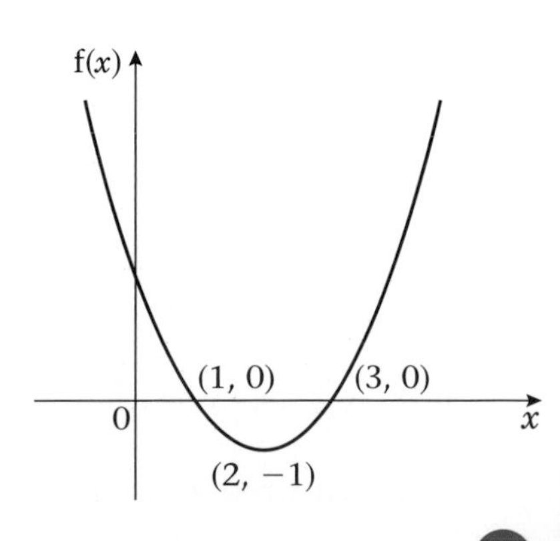

a Find the equation of the graph in the form $y = \mathrm{f}(x)$.

b On separate axes, sketch the graphs of
i $y = \mathrm{f}(x + 2)$ **ii** $y = \mathrm{f}(2x)$.

c On each graph write in the coordinates of the points at which the graph meets the x-axis and write in the coordinates of the minimum point.

E

Summary of key points

1 You should know the shapes of the following basic curves.

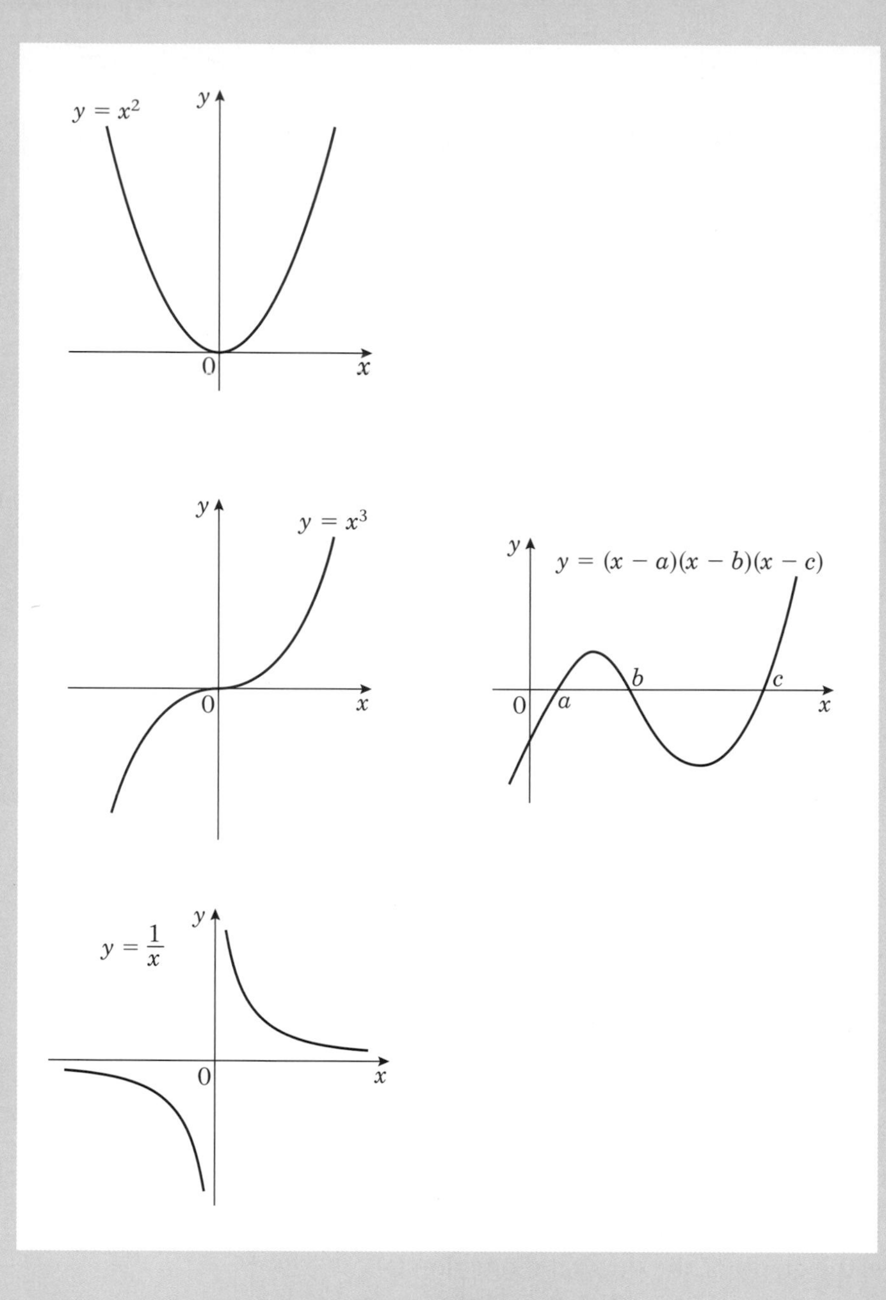

2 Transformations:

$f(x + a)$ is a translation of $-a$ in the x-direction.

$f(x) + a$ is a translation of $+a$ in the y-direction.

$f(ax)$ is a stretch of $\frac{1}{a}$ in the x-direction $\left(\text{multiply } x\text{-coordinates by } \frac{1}{a}\right)$.

$af(x)$ is a stretch of a in the y-direction (multiply y-coordinates by a).

1

Review Exercise

1 Factorise completely:

a $2x^3 - 13x^2 - 7x$

b $9x^2 - 16$

c $x^4 + 7x^2 - 8$.

2 Find the value of:

a $81^{\frac{1}{2}}$

b $81^{\frac{3}{4}}$

c $81^{-\frac{3}{4}}$. **E**

3 **a** Write down the value of $8^{\frac{1}{3}}$.

b Find the value of $8^{-\frac{2}{3}}$. **E**

4 **a** Find the value of $125^{\frac{4}{3}}$.

b Simplify $24x^2 \div 18x^{\frac{4}{3}}$.

5 **a** Express $\sqrt{80}$ in the form $a\sqrt{5}$, where a is an integer.

b Express $(4 - \sqrt{5})^2$ in the form $b + c\sqrt{5}$, where b and c are integers. **E**

6 **a** Expand and simplify $(4 + \sqrt{3})(4 - \sqrt{3})$.

b Express $\dfrac{26}{4 + \sqrt{3}}$ in the form $a + b\sqrt{3}$, where a and b are integers. **E**

7 **a** Express $\sqrt{108}$ in the form $a\sqrt{3}$, where a is an integer.

b Express $(2 - \sqrt{3})^2$ in the form $b + c\sqrt{3}$, where b and c are integers to be found. **E**

8 **a** Express $(2\sqrt{7})^3$ in the form $a\sqrt{7}$, where a is an integer.

b Express $(8 + \sqrt{7})(3 - 2\sqrt{7})$ in the form $b + c\sqrt{7}$, where b and c are integers.

c Express $\dfrac{6 + 2\sqrt{7}}{3 - \sqrt{7}}$ in the form $d + e\sqrt{7}$, where d and e are integers.

9 Solve the equations:

a $x^2 - x - 72 = 0$

b $2x^2 + 7x = 0$

c $10x^2 + 9x - 9 = 0$.

10 Solve the equations, giving your answers to 3 significant figures:

a $x^2 + 10x + 17 = 0$

b $2x^2 - 5x - 1 = 0$

c $(2x - 3)^2 = 7$.

11 $$x^2 - 8x - 29 \equiv (x + a)^2 + b,$$
where a and b are constants.

a Find the value of a and the value of b.

b Hence, or otherwise, show that the roots of
$$x^2 - 8x - 29 = 0$$
are $c \pm d\sqrt{5}$, where c and d are integers.

E

12 Given that
$$f(x) = x^2 - 6x + 18, x \geqslant 0,$$

a express f(x) in the form $(x - a)^2 + b$, where a and b are integers.

The curve C with equation $y = f(x)$, $x \geqslant 0$, meets the y-axis at P and has a minimum point at Q.

b Sketch the graph of C, showing the coordinates of P and Q.

The line $y = 41$ meets C at the point R.

c Find the x-coordinate of R, giving your answer in the form $p + q\sqrt{2}$, where p and q are integers.

E

13 Given that the equation $kx^2 + 12x + k = 0$, where k is a positive constant, has equal roots, find the value of k.

14 Given that
$$x^2 + 10x + 36 \equiv (x + a)^2 + b,$$
where a and b are constants,

a find the value of a and the value of b.

b Hence show that the equation $x^2 + 10x + 36 = 0$ has no real roots.

The equation $x^2 + 10x + k = 0$ has equal roots.

c Find the value of k.

d For this value of k, sketch the graph of $y = x^2 + 10x + k$, showing the coordinates of any points at which the graph meets the coordinate axes.

E

15 $$x^2 + 2x + 3 \equiv (x + a)^2 + b.$$

a Find the values of the constants a and b.

b Sketch the graph of $y = x^2 + 2x + 3$, indicating clearly the coordinates of any intersections with the coordinate axes.

c Find the value of the discriminant of $x^2 + 2x + 3$. Explain how the sign of the discriminant relates to your sketch in part **b**.

The equation $x^2 + kx + 3 = 0$, where k is a constant, has no real roots.

d Find the set of possible values of k, giving your answer in surd form.

E

16 Solve the simultaneous equations:
$$x + y = 2$$
$$x^2 + 2y = 12$$

17 **a** By eliminating y from the equations:
$$y = x - 4,$$
$$2x^2 - xy = 8,$$
show that
$$x^2 + 4x - 8 = 0.$$

b Hence, or otherwise, solve the simultaneous equations:
$$y = x - 4,$$
$$2x^2 - xy = 8,$$
giving your answers in the form $a \pm b\sqrt{3}$, where a and b are integers.

E

18 Solve the simultaneous equations:
$$2x - y - 5 = 0$$
$$x^2 + xy - 2 = 0$$

19 Find the set of values of x for which:

a $3(2x + 1) > 5 - 2x$,

b $2x^2 - 7x + 3 > 0$,

c both $3(2x + 1) > 5 - 2x$ and $2x^2 - 7x + 3 > 0$.

E

20 Find the set of values of x for which:

a $x(x - 5) < 7x - x^2$

b $x(3x + 7) > 20$

21 **a** Solve the simultaneous equations:

$$y + 2x = 5$$
$$2x^2 - 3x - y = 16.$$

b Hence, or otherwise, find the set of values of x for which:

$$2x^2 - 3x - 16 > 5 - 2x.$$

E

22 The equation $x^2 + kx + (k + 3) = 0$, where k is a constant, has different real roots.

a Show that $k^2 - 4k - 12 > 0$.

b Find the set of possible values of k. **E**

23 Given that the equation $kx^2 + 3kx + 2 = 0$, where k is a constant, has no real roots, find the set of possible values of k.

24 The equation $(2p + 5)x^2 + px + 1 = 0$, where p is a constant, has different real roots.

a Show that $p^2 - 8p - 20 > 0$.

b Find the set of possible values of p.

Given that $p = -3$,

c find the exact roots of $(2p + 5)x^2 + px + 1 = 0$.

25 **a** Factorise completely $x^3 - 4x$.

b Sketch the curve with equation $y = x^3 - 4x$, showing the coordinates of the points where the curve crosses the x-axis.

c On a separate diagram, sketch the curve with equation

$$y = (x - 1)^3 - 4(x - 1)$$

showing the coordinates of the points where the curve crosses the x-axis. **E**

26

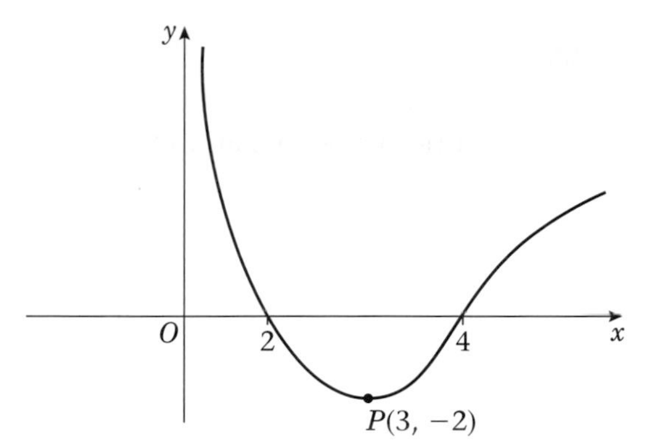

The figure shows a sketch of the curve with equation $y = f(x)$. The curve crosses the x-axis at the points $(2, 0)$ and $(4, 0)$. The minimum point on the curve is $P(3, -2)$.

In separate diagrams, sketch the curve with equation

a $y = -f(x)$ **b** $y = f(2x)$

On each diagram, give the coordinates of the points at which the curve crosses the x-axis, and the coordinates of the image of P under the given transformation. **E**

27

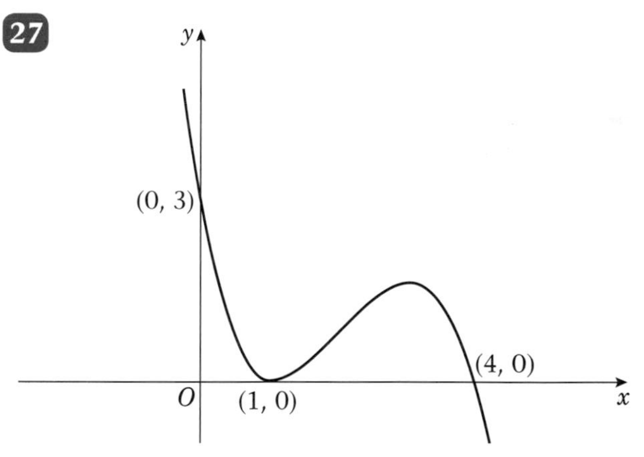

The figure shows a sketch of the curve with equation $y = f(x)$. The curve passes through the points $(0, 3)$ and $(4, 0)$ and touches the x-axis at the point $(1, 0)$.

On separate diagrams, sketch the curve with equation

a $y = f(x + 1)$ **b** $y = 2f(x)$

c $y = f(\frac{1}{2}x)$

On each diagram, show clearly the coordinates of all the points where the curve meets the axes. **E**

28 Given that $f(x) = \frac{1}{x}, x \neq 0$,

a sketch the graph of $y = f(x) + 3$ and state the equations of the asymptotes

b find the coordinates of the point where $y = f(x) + 3$ crosses a coordinate axis. *E*

29 Given that $f(x) = (x^2 - 6x)(x - 2) + 3x$,

a express $f(x)$ in the form $x(ax^2 + bx + c)$, where a, b and c are constants

b hence factorise $f(x)$ completely

c sketch the graph of $y = f(x)$, showing the coordinates of each point at which the graph meets the axes.

30 **a** Sketch on the same diagram the graph of $y = x(x + 2)(x - 4)$ and the graph of $y = 3x - x^2$, showing the coordinates of the points at which each graph meets the x-axis.

b Find the exact coordinates of each of the intersection points of $y = x(x + 2)(x - 4)$ and $y = 3x - x^2$.

After completing this chapter you should be able to

1. understand the link between the equation of a line, and its gradient and intercept
2. calculate the gradient of a line joining a pair of points
3. find the equation of a line in either the form $y = mx + c$ or alternatively $ax + by = c$
4. find the equation of a line passing through a pair of points
5. determine the point where a pair of straight lines intersect
6. know and use the rule concerning perpendicular gradients.

Understanding this chapter will help you find the equation of a tangent and normal to a curve in Chapter 7.

Coordinate geometry in the (x, y) plane

Did you know?

...that many bills (including mobile phones) are linear and will produce straight lines when they are graphed?

The problem below can easily be answered by solving where pair of straight lines intersect.
If C = the cost of the calls in £s and t = the time in minutes, then the graphs are

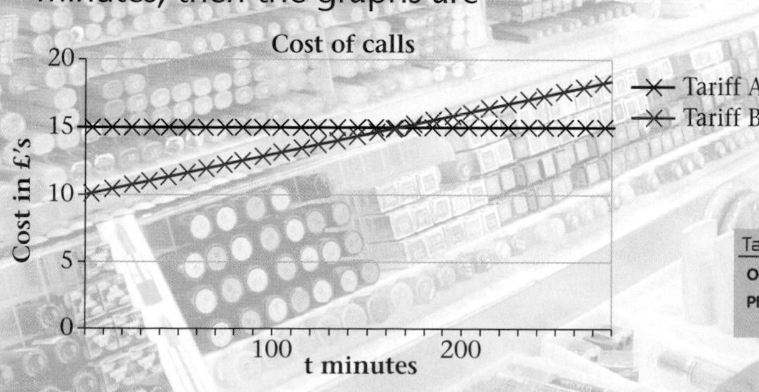

Tariff A
£15 per month
ONLY!

Tariff B | A STANDING CHARGE OF £10 PER MONTH PLUS 3p PER MINUTE FOR CALLS

Can you work out when it would be cheaper to use Tariff A?
You may wish to work out if your mobile phone contract is the most suitable one for the number of calls you make.

5.1 You can write the equation of a straight line in the form $y = mx + c$ or $ax + by + c = 0$.

- **In the general form $y = mx + c$, m is the gradient and $(0, c)$ is the intercept on the y-axis.**

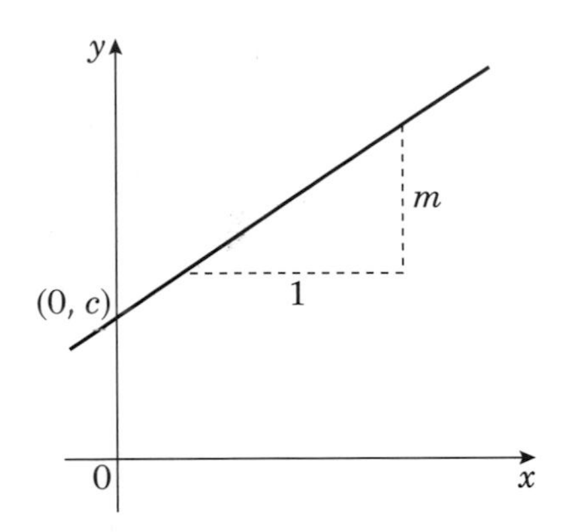

- **In the general form $ax + by + c = 0$, a, b and c are integers.**

Example 1

Write down the gradient and intercept on the y-axis of these lines:

a $y = -3x + 2$

b $4x - 2y + 5 = 0$

a $y = -3x + 2$

The gradient $= -3$ and the intercept on the y-axis $= (0, 2)$.

Compare $y = -3x + 2$ with $y = mx + c$.
From this, $m = -3$ and $c = 2$.

b $4x - 2y + 5 = 0$

Rearrange the equation into the form $y = mx + c$.

$4x + 5 = 2y$

Add $2y$ to each side.

So $2y = 4x + 5$

Put the term in y at the front of the equation.

$y = 2x + \frac{5}{2}$

Divide each term by 2, so that:

$2y \div 2 = y$
$4 \div 2 = 2$
$5 \div 2 = \frac{5}{2}$. (Do not write this as 2.5)

The gradient $= 2$ and the intercept on the y-axis $= (0, \frac{5}{2})$.

Compare $y = 2x + \frac{5}{2}$ to $y = mx + c$.
From this, $m = 2$ and $c = \frac{5}{2}$.

Example 2

Write these lines in the form $ax + by + c = 0$:

a $y = 4x + 3$ **b** $y = -\frac{1}{2}x + 5$

a $y = 4x + 3$

$0 = 4x + 3 - y$

So $4x - y + 3 = 0$

b $y = -\frac{1}{2}x + 5$

$\frac{1}{2}x + y = 5$

$\frac{1}{2}x + y - 5 = 0$

So $x + 2y - 10 = 0$

Rearrange the equation into the form $ax + by + c = 0$.

Subtract y from each side.

Collect all the terms on one side of the equation.

Add $\frac{1}{2}x$ to each side.

Subtract 5 from each side.

Multiply each term by 2 to clear the fraction.

Example 3

A line is parallel to the line $y = \frac{1}{2}x - 5$ and its intercept on the y-axis is (0, 1). Write down the equation of the line.

$y = \frac{1}{2}x + 1$

Remember that parallel lines have the same gradient.

Compare $y = \frac{1}{2}x - 5$ with $y = mx + c$, so $m = \frac{1}{2}$.

The gradient of the required line $= \frac{1}{2}$.

The intercept on the y-axis is (0, 1), so $c = 1$.

Example 4

A line is parallel to the line $6x + 3y - 2 = 0$ and it passes through the point (0, 3). Work out the equation of the line.

$6x + 3y - 2 = 0$

$3y - 2 = -6x$

$3y = -6x + 2$

$y = -2x + \frac{2}{3}$

The gradient of this line is -2.

The equation of the line is $y = -2x + 3$.

Rearrange the equation into the form $y = mx + c$ to find m.

Subtract $6x$ from each side.

Add 2 to each side.

Divide each term by 3, so that

$3y \div 3 = y$

$-6x \div 3 = -2x$

$2 \div 3 = \frac{2}{3}$. (Do not write this as a decimal.)

Compare $y = -2x + \frac{2}{3}$ with $y = mx + c$, so $m = -2$.

Parallel lines have the same gradient, so the gradient of the required line $= -2$.

(0, 3) is the intercept on the y-axis, so $c = 3$.

Example 5

The line $y = 4x - 8$ meets the x-axis at the point P. Work out the coordinates of P.

$y = 4x - 8$

Substituting,

$4x - 8 = 0$

$4x = 8$ — Add 8 to each side.

$x = 2$ — Divide each side by 4.

So $P(2, 0)$.

The line meets the x-axis when $y = 0$, so substitute $y = 0$ into $y = 4x - 8$.

Rearrange the equation for x.

Always write down the coordinates of the point.

Exercise 5A

1 Work out the gradients of these lines:

a $y = -2x + 5$ **b** $y = -x + 7$ **c** $y = 4 + 3x$

d $y = \frac{1}{3}x - 2$ **e** $y = -\frac{2}{3}x$ **f** $y = \frac{5}{4}x + \frac{2}{3}$

g $2x - 4y + 5 = 0$ **h** $10x - 5y + 1 = 0$ **i** $-x + 2y - 4 = 0$

j $-3x + 6y + 7 = 0$ **k** $4x + 2y - 9 = 0$ **l** $9x + 6y + 2 = 0$

2 These lines intercept the y-axis at $(0, c)$. Work out the value of c in each case.

a $y = -x + 4$ **b** $y = 2x - 5$ **c** $y = \frac{1}{2}x - \frac{2}{3}$

d $y = -3x$ **e** $y = \frac{6}{7}x + \frac{7}{5}$ **f** $y = 2 - 7x$

g $3x - 4y + 8 = 0$ **h** $4x - 5y - 10 = 0$ **i** $-2x + y - 9 = 0$

j $7x + 4y + 12 = 0$ **k** $7x - 2y + 3 = 0$ **l** $-5x + 4y + 2 = 0$

3 Write these lines in the form $ax + by + c = 0$.

a $y = 4x + 3$ **b** $y = 3x - 2$ **c** $y = -6x + 7$

d $y = \frac{4}{5}x - 6$ **e** $y = \frac{5}{3}x + 2$ **f** $y = \frac{7}{3}x$

g $y = 2x - \frac{4}{7}$ **h** $y = -3x + \frac{2}{9}$ **i** $y = -6x - \frac{2}{3}$

j $y = -\frac{1}{3}x + \frac{1}{2}$ **k** $y = \frac{2}{3}x + \frac{5}{6}$ **l** $y = \frac{3}{5}x + \frac{1}{2}$

4 A line is parallel to the line $y = 5x + 8$ and its intercept on the y-axis is $(0, 3)$. Write down the equation of the line.

5 A line is parallel to the line $y = -\frac{2}{5}x + 1$ and its intercept on the y-axis is $(0, -4)$. Work out the equation of the line. Write your answer in the form $ax + by + c = 0$, where a, b and c are integers.

6 A line is parallel to the line $3x + 6y + 11 = 0$ and its intercept on the y-axis is $(0, 7)$. Write down the equation of the line.

7 A line is parallel to the line $2x - 3y - 1 = 0$ and it passes through the point $(0, 0)$. Write down the equation of the line.

8 The line $y = 6x - 18$ meets the x-axis at the point P. Work out the coordinates of P.

9 The line $3x + 2y - 5 = 0$ meets the x-axis at the point R. Work out the coordinates of R.

10 The line $5x - 4y + 20 = 0$ meets the y-axis at the point A and the x-axis at the point B. Work out the coordinates of the points A and B.

5.2 You can work out the gradient m of the line joining the point with coordinates (x_1, y_1) to the point with coordinates (x_2, y_2) by using the formula $m = \frac{y_2 - y_1}{x_2 - x_1}$.

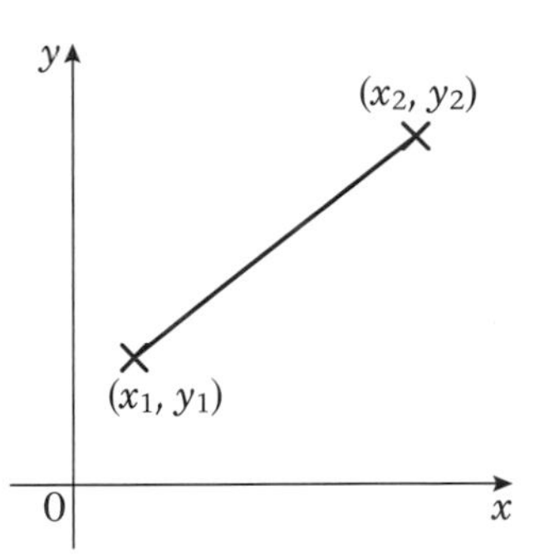

Example 6

Work out the gradient of the line joining the points (2, 3) and (5, 7).

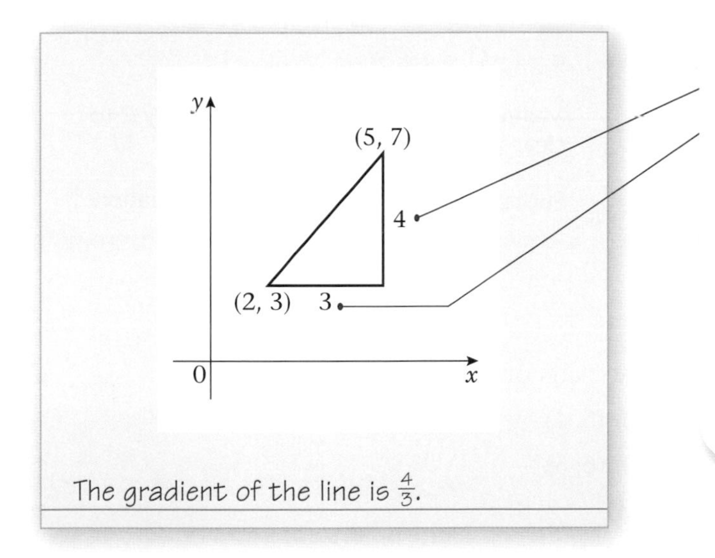

Draw a sketch.

$7 - 3 = 4$

$5 - 2 = 3$

Remember the gradient of a line $= \frac{\text{difference in } y\text{-coordinates}}{\text{difference in } x\text{-coordinates}}$,

so $m = \frac{7 - 3}{5 - 2}$.

This is $m = \frac{y_2 - y_1}{x_2 - x_1}$ with $(x_1, y_1) = (2, 3)$ and $(x_2, y_2) = (5, 7)$.

The gradient of the line is $\frac{4}{3}$.

Example 7

Work out the gradient of the line joining these pairs of points:

a $(-2, 7)$ and $(4, 5)$ **b** $(2d, -5d)$ and $(6d, 3d)$

a $m = \frac{5 - 7}{4 - (-2)}$

Use $m = \frac{y_2 - y_1}{x_2 - x_1}$. Here $(x_1, y_1) = (-2, 7)$ and $(x_2, y_2) = (4, 5)$.

$= \frac{-2}{6}$

$-(-2) = +2$, so $4 + 2 = 6$

$= -\frac{1}{3}$

Remember to simplify the fraction when possible, so divide by 2.

The gradient of the line is $-\frac{1}{3}$.

$\frac{-1}{3}$ is the same as $-\frac{1}{3}$.

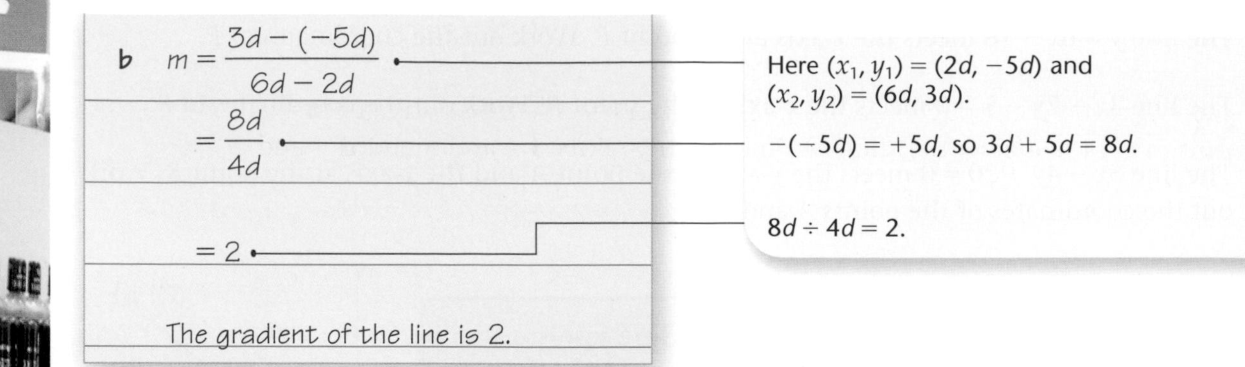

b $m = \frac{3d - (-5d)}{6d - 2d}$ — Here $(x_1, y_1) = (2d, -5d)$ and $(x_2, y_2) = (6d, 3d)$.

$= \frac{8d}{4d}$ — $-(-5d) = +5d$, so $3d + 5d = 8d$.

$= 2$ — $8d \div 4d = 2$.

The gradient of the line is 2.

Example 8

The line joining $(2, -5)$ to $(4, a)$ has gradient -1. Work out the value of a.

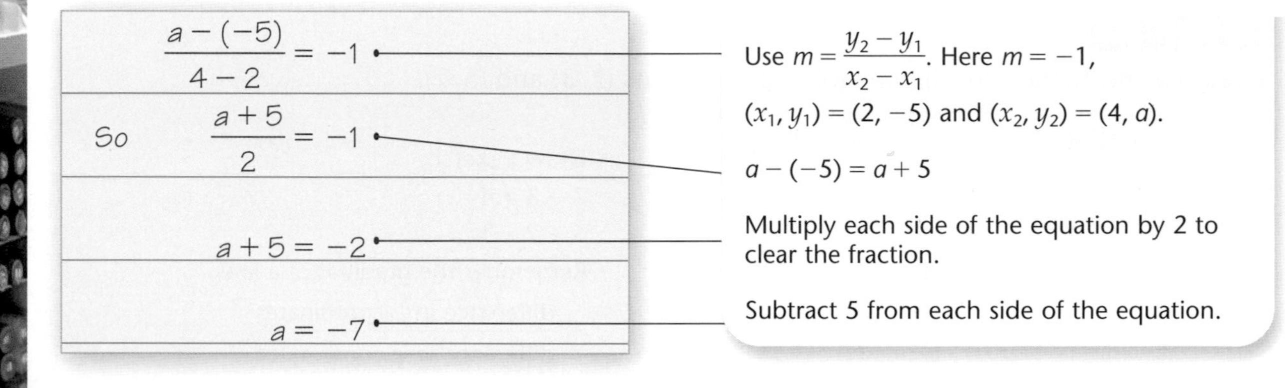

$\frac{a - (-5)}{4 - 2} = -1$ — Use $m = \frac{y_2 - y_1}{x_2 - x_1}$. Here $m = -1$, $(x_1, y_1) = (2, -5)$ and $(x_2, y_2) = (4, a)$.

So $\frac{a + 5}{2} = -1$ — $a - (-5) = a + 5$

$a + 5 = -2$ — Multiply each side of the equation by 2 to clear the fraction.

$a = -7$ — Subtract 5 from each side of the equation.

Exercise 5B

1 Work out the gradient of the line joining these pairs of points:

a $(4, 2), (6, 3)$ **b** $(-1, 3), (5, 4)$

c $(-4, 5), (1, 2)$ **d** $(2, -3), (6, 5)$

e $(-3, 4), (7, -6)$ **f** $(-12, 3), (-2, 8)$

g $(-2, -4), (10, 2)$ **h** $(\frac{1}{2}, 2), (\frac{3}{4}, 4)$

i $(\frac{1}{4}, \frac{1}{2}), (\frac{1}{2}, \frac{2}{3})$ **j** $(-2.4, 9.6), (0, 0)$

k $(1.3, -2.2), (8.8, -4.7)$ **l** $(0, 5a), (10a, 0)$

m $(3b, -2b), (7b, 2b)$ **n** $(p, p^2), (q, q^2)$

2 The line joining $(3, -5)$ to $(6, a)$ has gradient 4. Work out the value of a.

3 The line joining $(5, b)$ to $(8, 3)$ has gradient -3. Work out the value of b.

4 The line joining $(c, 4)$ to $(7, 6)$ has gradient $\frac{3}{4}$. Work out the value of c.

5 The line joining $(-1, 2d)$ to $(1, 4)$ has gradient $-\frac{1}{4}$. Work out the value of d.

6 The line joining $(-3, -2)$ to $(2e, 5)$ has gradient 2. Work out the value of e.

7 The line joining $(7, 2)$ to $(f, 3f)$ has gradient 4. Work out the value of f.

8 The line joining $(3, -4)$ to $(-g, 2g)$ has gradient -3. Work out the value of g.

9 Show that the points $A(2, 3)$, $B(4, 4)$, $C(10, 7)$ can be joined by a straight line.
(Hint: Find the gradient of the lines joining the points: **i** A and B and **ii** A and C.)

10 Show that the points $(-2a, 5a)$, $(0, 4a)$, $(6a, a)$ are collinear (i.e. on the same straight line).

5.3 **You can find the equation of a line with gradient m that passes through the point with coordinates (x_1, y_1) by using the formula $y - y_1 = m(x - x_1)$.**

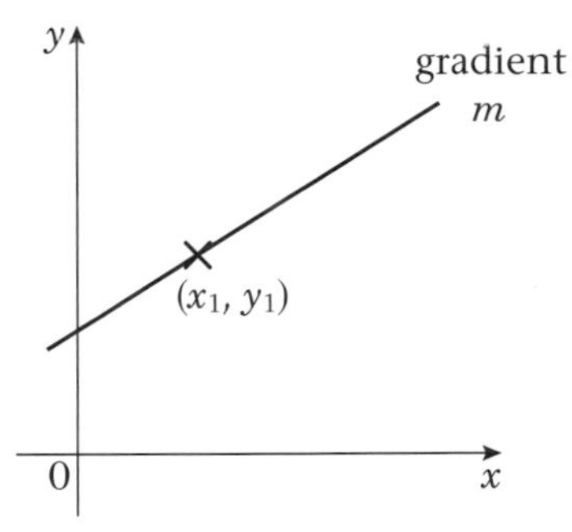

Example 9

Find the equation of the line with gradient 5 that passes through the point (3, 2).

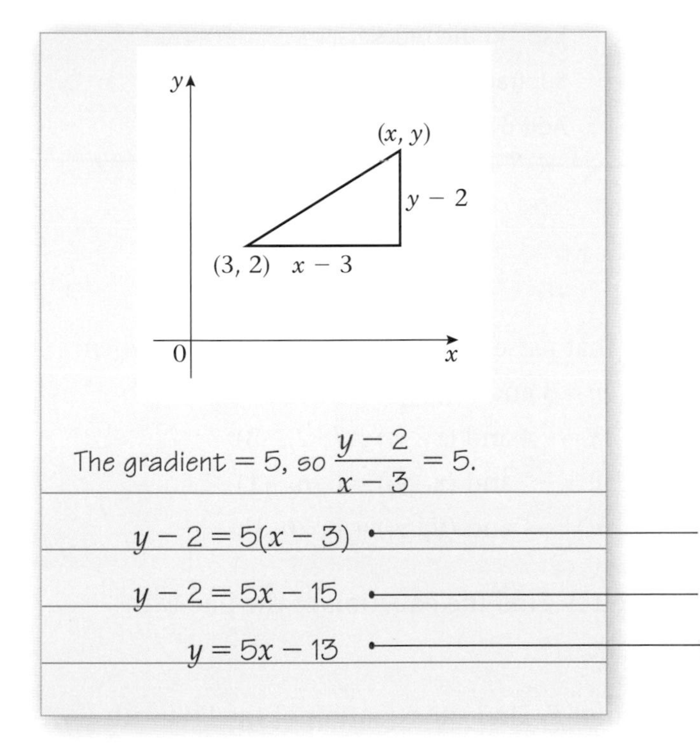

(x, y) is *any* point on the line.

The gradient = 5, so $\dfrac{y-2}{x-3} = 5$.

Multiply each side of the equation by $x - 3$ to clear the fraction, so that:

$$\frac{y-2}{x-3} \times \frac{x-3}{1} = y - 2$$

$$5 \times (x - 3) = 5(x - 3)$$

$y - 2 = 5(x - 3)$ — This is in the form $y - y_1 = m(x - x_1)$. Here $m = 5$ and $(x_1, y_1) = (3, 2)$.

$y - 2 = 5x - 15$ — Expand the brackets.

$y = 5x - 13$ — Add 2 to each side.

Example 10

Find the equation of the line with gradient $-\frac{1}{2}$ that passes through the point $(4, -6)$.

$y - (-6) = -\frac{1}{2}(x - 4)$ — Use $y - y_1 = m(x - x_1)$. Here $m = -\frac{1}{2}$ and $(x_1, y_1) = (4, -6)$.

So $y + 6 = -\frac{1}{2}(x - 4)$

$y + 6 = -\frac{1}{2}x + 2$ — Expand the brackets. Remember $-\frac{1}{2} \times -4 = +2$.

$y = -\frac{1}{2}x - 4$ — Subtract 6 from each side.

Example 11

The line $y = 3x - 9$ meets the x-axis at the point A. Find the equation of the line with gradient $\frac{2}{3}$ that passes through the point A. Write your answer in the form $ax + by + c = 0$, where a, b and c are integers.

$y = 3x - 9$

$3x - 9 = 0$ — The line meets the x-axis when $y = 0$, so substitute $y = 0$ into $y = 3x - 9$.

$3x = 9$ — Rearrange the equation to find x.

$x = 3$

So $A(3, 0)$. — Always write down the coordinates of the point.

$y - 0 = \frac{2}{3}(x - 3)$ — Use $y - y_1 = m(x - x_1)$. Here $m = \frac{2}{3}$ and $(x_1, y_1) = (3, 0)$.

$y = \frac{2}{3}(x - 3)$ — Rearrange the equation into the form $ax + by + c = 0$.

$3y = 2(x - 3)$ — Multiply by 3 to clear the fraction.

$3y = 2x - 6$ — Expand the brackets.

$-2x + 3y = -6$ — Subtract $2x$ from each side.

$-2x + 3y + 6 = 0$ — Add 6 to each side.

Exercise 5C

1 Find the equation of the line with gradient m that passes through the point (x_1, y_1) when:

a $m = 2$ and $(x_1, y_1) = (2, 5)$

b $m = 3$ and $(x_1, y_1) = (-2, 1)$

c $m = -1$ and $(x_1, y_1) = (3, -6)$

d $m = -4$ and $(x_1, y_1) = (-2, -3)$

e $m = \frac{1}{2}$ and $(x_1, y_1) = (-4, 10)$

f $m = -\frac{2}{3}$ and $(x_1, y_1) = (-6, -1)$

g $m = 2$ and $(x_1, y_1) = (a, 2a)$

h $m = -\frac{1}{2}$ and $(x_1, y_1) = (-2b, 3b)$

2 The line $y = 4x - 8$ meets the x-axis at the point A. Find the equation of the line with gradient 3 that passes through the point A.

3 The line $y = -2x + 8$ meets the y-axis at the point B. Find the equation of the line with gradient 2 that passes through the point B.

4 The line $y = \frac{1}{2}x + 6$ meets the x-axis at the point C. Find the equation of the line with gradient $\frac{2}{3}$ that passes through the point C. Write your answer in the form $ax + by + c = 0$, where a, b and c are integers.

5 The line $y = \frac{1}{4}x + 2$ meets the y-axis at the point B. The point C has coordinates $(-5, 3)$. Find the gradient of the line joining the points B and C.

6 The lines $y = x$ and $y = 2x - 5$ intersect at the point A. Find the equation of the line with gradient $\frac{2}{5}$ that passes through the point A. (Hint: Solve $y = x$ and $y = 2x - 5$ simultaneously.)

7 The lines $y = 4x - 10$ and $y = x - 1$ intersect at the point T. Find the equation of the line with gradient $-\frac{2}{3}$ that passes through the point T. Write your answer in the form $ax + by + c = 0$, where a, b and c are integers.

8 The line p has gradient $\frac{2}{3}$ and passes through the point $(6, -12)$. The line q has gradient -1 and passes through the point $(5, 5)$. The line p meets the y-axis at A and the line q meets the x-axis at B. Work out the gradient of the line joining the points A and B.

9 The line $y = -2x + 6$ meets the x-axis at the point P. The line $y = \frac{3}{2}x - 4$ meets the y-axis at the point Q. Find the equation of the line joining the points P and Q. (Hint: First work out the gradient of the line joining the points P and Q.)

10 The line $y = 3x - 5$ meets the x-axis at the point M. The line $y = -\frac{2}{3}x + \frac{2}{3}$ meets the y-axis at the point N. Find the equation of the line joining the points M and N. Write your answer in the form $ax + by + c = 0$, where a, b and c are integers.

5.4 You can find the equation of the line that passes through the points with coordinates (x_1, y_1) and (x_2, y_2) by using the formula $\dfrac{y - y_1}{y_2 - y_1} = \dfrac{x - x_1}{x_2 - x_1}$.

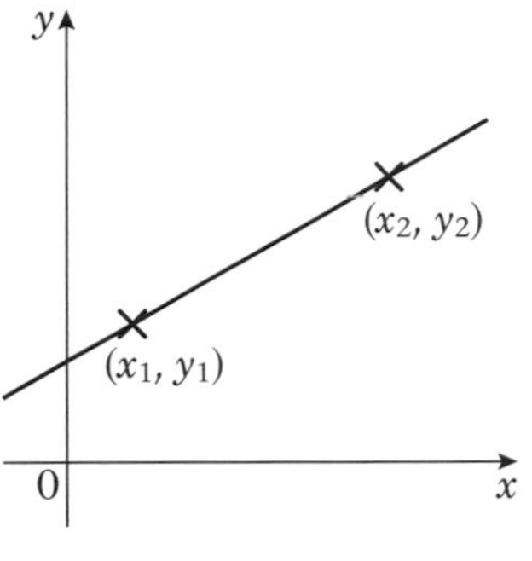

Example 12

Work out the gradient of the line that passes through the points $(5, 7)$ and $(3,\;\; 1)$ and hence find the equation of the line.

$$m = \frac{(-1) - 7}{3 - 5}$$

Use $m = \dfrac{y_2 - y_1}{x_2 - x_1}$. Here $(x_1, y_1) = (5, 7)$ and $(x_2, y_2) = (3, -1)$.

$$= \frac{-8}{-2}$$

$-8 \div -2 = +4$

So $m = 4$.

$$y - 7 = 4(x - 5)$$

Use $y - y_1 = m(x - x_1)$. Here $m = 4$ and $(x_1, y_1) = (5, 7)$.

$$y - 7 = 4x - 20$$

Expand the brackets.

$$y = 4x - 13$$

Simplify into the form $y = mx + c$. Add 7 to each side.

Example 13

Use $\dfrac{y - y_1}{y_2 - y_1} = \dfrac{x - x_1}{x_2 - x_1}$ to find the equation of the line that passes through the points $(5, 7)$ and $(3, -1)$.

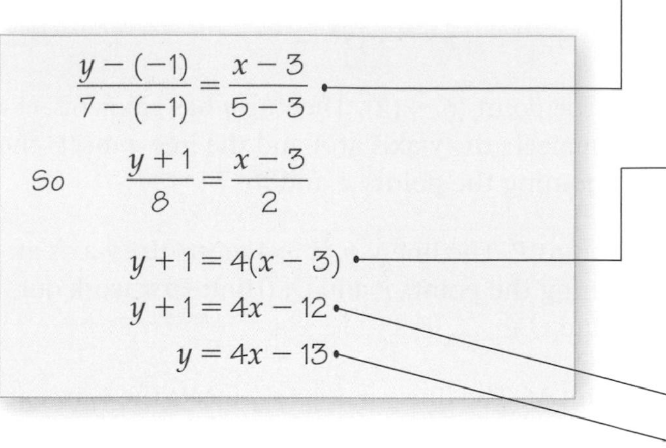

$$\frac{y-(-1)}{7-(-1)} = \frac{x-3}{5-3}$$

Use $\dfrac{y - y_1}{y_2 - y_1} = \dfrac{x - x_1}{x_2 - x_1}$.

Here $(x_1, y_1) = (3, -1)$ and $(x_2, y_2) = (5, 7)$.

(x_1, y_1) and (x_2, y_2) have been chosen to make the denominators positive.

So $\dfrac{y+1}{8} = \dfrac{x-3}{2}$

$y + 1 = 4(x - 3)$

Multiply each side by 8 to clear the fraction, so that:

$$8 \times \frac{y+1}{8} = y + 1$$

$$8 \times \frac{x-3}{2} = 4(x-3)$$

$y + 1 = 4x - 12$ — Expand the brackets.

$y = 4x - 13$ — Subtract 1 from each side.

Example 14

The lines $y = 4x - 7$ and $2x + 3y - 21 = 0$ intersect at the point A. The point B has coordinates $(-2, 8)$. Find the equation of the line that passes through the points A and B. Write your answer in the form $ax + by + c = 0$, where a, b and c are integers.

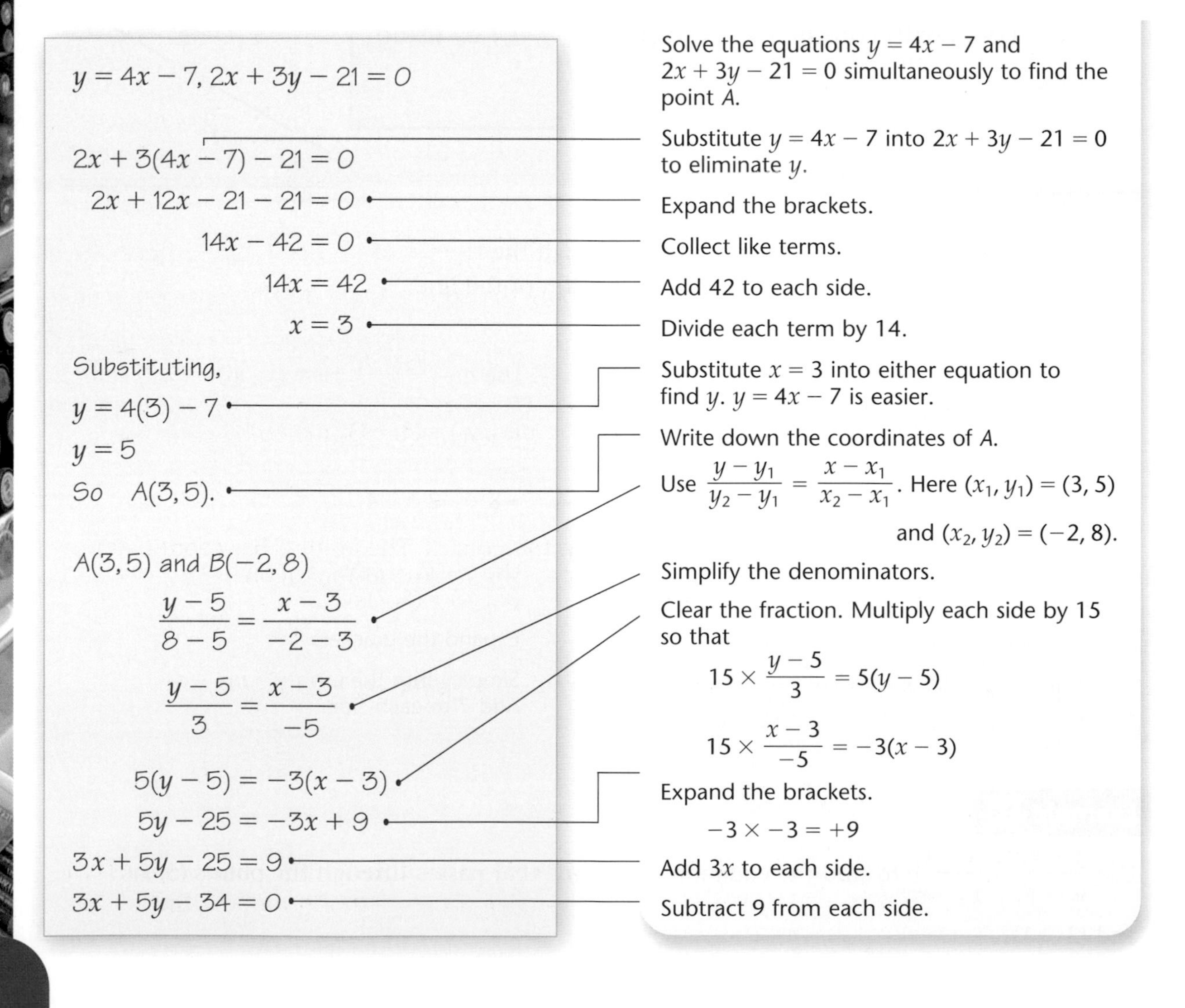

$y = 4x - 7,\ 2x + 3y - 21 = 0$ — Solve the equations $y = 4x - 7$ and $2x + 3y - 21 = 0$ simultaneously to find the point A.

$2x + 3(4x - 7) - 21 = 0$ — Substitute $y = 4x - 7$ into $2x + 3y - 21 = 0$ to eliminate y.

$2x + 12x - 21 - 21 = 0$ — Expand the brackets.

$14x - 42 = 0$ — Collect like terms.

$14x = 42$ — Add 42 to each side.

$x = 3$ — Divide each term by 14.

Substituting,

$y = 4(3) - 7$ — Substitute $x = 3$ into either equation to find y. $y = 4x - 7$ is easier.

$y = 5$

So $A(3, 5)$. — Write down the coordinates of A.

$A(3, 5)$ and $B(-2, 8)$

$\dfrac{y-5}{8-5} = \dfrac{x-3}{-2-3}$ — Use $\dfrac{y - y_1}{y_2 - y_1} = \dfrac{x - x_1}{x_2 - x_1}$. Here $(x_1, y_1) = (3, 5)$ and $(x_2, y_2) = (-2, 8)$.

$\dfrac{y-5}{3} = \dfrac{x-3}{-5}$ — Simplify the denominators.

$5(y - 5) = -3(x - 3)$ — Clear the fraction. Multiply each side by 15 so that

$$15 \times \frac{y-5}{3} = 5(y-5)$$

$$15 \times \frac{x-3}{-5} = -3(x-3)$$

$5y - 25 = -3x + 9$ — Expand the brackets. $-3 \times -3 = +9$

$3x + 5y - 25 = 9$ — Add $3x$ to each side.

$3x + 5y - 34 = 0$ — Subtract 9 from each side.

Exercise 5D

1 Find the equation of the line that passes through these pairs of points:

a $(2, 4)$ and $(3, 8)$

b $(0, 2)$ and $(3, 5)$

c $(-2, 0)$ and $(2, 8)$

d $(5, -3)$ and $(7, 5)$

e $(3, -1)$ and $(7, 3)$

f $(-4, -1)$ and $(6, 4)$

g $(-1, -5)$ and $(-3, 3)$

h $(-4, -1)$ and $(-3, -9)$

i $(\frac{1}{3}, \frac{2}{5})$ and $(\frac{2}{3}, \frac{4}{5})$

j $(-\frac{3}{4}, \frac{1}{7})$ and $(\frac{1}{4}, \frac{3}{7})$

2 The line that passes through the points $(2, -5)$ and $(-7, 4)$ meets the x-axis at the point P. Work out the coordinates of the point P.

3 The line that passes through the points $(-3, -5)$ and $(4, 9)$ meets the y-axis at the point G. Work out the coordinates of the point G.

4 The line that passes through the points $(3, 2\frac{1}{2})$ and $(-1\frac{1}{2}, 4)$ meets the y-axis at the point J. Work out the coordinates of the point J.

5 The line $y = 2x - 10$ meets the x-axis at the point A. The line $y = -2x + 4$ meets the y-axis at the point B. Find the equation of the line joining the points A and B. (Hint: First work out the coordinates of the points A and B.)

6 The line $y = 4x + 5$ meets the y-axis at the point C. The line $y = -3x - 15$ meets the x-axis at the point D. Find the equation of the line joining the points C and D. Write your answer in the form $ax + by + c = 0$, where a, b and c are integers.

7 The lines $y = x - 5$ and $y = 3x - 13$ intersect at the point S. The point T has coordinates $(-4, 2)$. Find the equation of the line that passes through the points S and T.

8 The lines $y = -2x + 1$ and $y = x + 7$ intersect at the point L. The point M has coordinates $(-3, 1)$. Find the equation of the line that passes through the points L and M.

9 The vertices of the triangle ABC have coordinates $A(3, 5)$, $B(-2, 0)$ and $C(4, -1)$. Find the equations of the sides of the triangle.

10 The line V passes through the points $(-5, 3)$ and $(7, -3)$ and the line W passes through the points $(2, -4)$ and $(4, 2)$. The lines V and W intersect at the point A. Work out the coordinates of the point A.

5.5 You can work out the gradient of a line that is perpendicular to the line $y = mx + c$.

- If a line has a gradient of m, a line perpendicular to it has a gradient of $-\frac{1}{m}$.
- If two lines are perpendicular, the product of their gradients is -1.

Example 15

Work out the gradient of the line that is perpendicular to the lines with these gradients:

a 3 **b** $\frac{1}{2}$ **c** $-\frac{2}{5}$

a $m = 3$

So the gradient of the perpendicular line is $-\frac{1}{3}$.

Use $-\frac{1}{m}$ with $m = 3$.

b $m = \frac{1}{2}$

So the gradient of the perpendicular line is

$$-\frac{1}{\left(\frac{1}{2}\right)}$$

$$= -\frac{2}{1}$$

$$= -2$$

Use $-\frac{1}{m}$ with $m = \frac{1}{2}$.

Remember $\frac{1}{\left(\frac{a}{b}\right)} = \frac{b}{a}$, so $\frac{1}{\left(\frac{1}{2}\right)} = \frac{2}{1}$.

c $m = -\frac{2}{5}$

So the gradient of the perpendicular line is

$$-\frac{1}{\left(-\frac{2}{5}\right)}$$

$$= -\left(-\frac{5}{2}\right)$$

$$= \frac{5}{2}$$

Use $-\frac{1}{m}$ with $m = -\frac{2}{5}$.

Here $\frac{1}{\left(\frac{2}{5}\right)} = \frac{5}{2}$, so $\frac{1}{\left(-\frac{2}{5}\right)} = -\frac{5}{2}$.

$-1 \times -\frac{5}{2} = +\frac{5}{2}$

Example 16

Show that the line $y = 3x + 4$ is perpendicular to the line $x + 3y - 3 = 0$.

$y = 3x + 4$

The gradient of this line is 3.

Compare $y = 3x + 4$ with $y = mx + c$, so $m = 3$.

$x + 3y - 3 = 0$

Rearrange the equation into the form $y = mx + c$ to find m.

$3y - 3 = -x$ — Subtract x from each side.

$3y = -x + 3$ — Add 3 to each side.

$y = -\frac{1}{3}x + 1$ — Divide each term by 3. $-x \div 3 = \frac{-x}{3} = -\frac{1}{3}x$.

The gradient of this line is $-\frac{1}{3}$. — Compare $y = -\frac{1}{3}x + 1$ with $y = mx + c$, so $m = -\frac{1}{3}$.

$3 \times -\frac{1}{3} = -1$ — Multiply the gradients of the lines.

The lines are perpendicular because the product of their gradients is -1.

Example 17

Work out whether these pairs of lines are parallel, perpendicular or neither:

a $y = -2x + 9$
$y = -2x - 3$

b $3x - y - 2 = 0$
$x + 3y - 6 = 0$

c $y = \frac{1}{2}x$
$2x - y + 4 = 0$

a $y = -2x + 9$

The gradient of this line is -2. — Compare $y = -2x + 9$ with $y = mx + c$, so $m = -2$.

$y = -2x - 3$ — Compare $y = -2x - 3$ with $y = mx + c$, so $m = -2$.

The gradient of this line is -2.

So the lines are parallel, since the gradients are equal. — Remember that parallel lines have the same gradient.

b $3x - y - 2 = 0$ — Rearrange the equation into the form $y = mx + c$.

$3x - 2 = y$ — Add y to each side.

So $y = 3x - 2$

The gradient of this line is 3. — Compare $y = 3x - 2$ with $y = mx + c$, so $m = 3$.

$x + 3y - 6 = 0$

$3y - 6 = -x$ — Subtract x from each side.

$3y = -x + 6$ — Add 6 to each side.

$y = -\frac{1}{3}x + 2$ — Divide each term by 3.

The gradient of this line is $-\frac{1}{3}$. — Compare $y = -\frac{1}{3}x + 2$ with $y = mx + c$, so $m = -\frac{1}{3}$.

So the lines are perpendicular as $3 \times \frac{1}{3} = -1$.

c $y = \frac{1}{2}x$ — Compare $y = \frac{1}{2}x$ with $y = mx + c$, so $m = \frac{1}{2}$.

The gradient of this line is $\frac{1}{2}$.

$2x - y + 4 = 0$ — Rearrange the equation into the form $y = mx + c$ to find m.

$2x + 4 = y$ — Add y to each side.

So $y = 2x + 4$

The gradient of this line is 2. — Compare $y = 2x + 4$ with $y = mx + c$, so $m = 2$.

The lines are not parallel as they have different gradients.

The lines are not perpendicular as $\frac{1}{2} \times 2 = 1$.

Example 18

Find an equation of the line that passes through the point $(3, -1)$ and is perpendicular to the line $y = 2x - 4$.

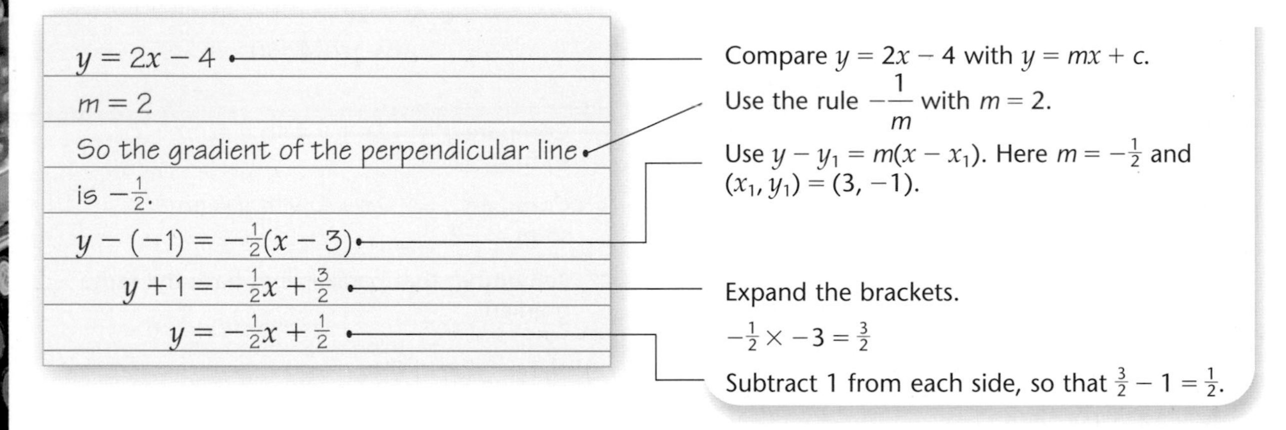

$y = 2x - 4$ — Compare $y = 2x - 4$ with $y = mx + c$.

$m = 2$

So the gradient of the perpendicular line is $-\frac{1}{2}$. — Use the rule $-\dfrac{1}{m}$ with $m = 2$.

$y - (-1) = -\frac{1}{2}(x - 3)$ — Use $y - y_1 = m(x - x_1)$. Here $m = -\frac{1}{2}$ and $(x_1, y_1) = (3, -1)$.

$y + 1 = -\frac{1}{2}x + \frac{3}{2}$ — Expand the brackets. $-\frac{1}{2} \times -3 = \frac{3}{2}$

$y = -\frac{1}{2}x + \frac{1}{2}$ — Subtract 1 from each side, so that $\frac{3}{2} - 1 = \frac{1}{2}$.

Exercise 5E

1 Work out whether these pairs of lines are parallel, perpendicular or neither:

a $y = 4x + 2$, $y = -\frac{1}{4}x - 7$

b $y = \frac{2}{3}x - 1$, $y = \frac{2}{3}x - 11$

c $y = \frac{1}{5}x + 9$, $y = 5x + 9$

d $y = -3x + 2$, $y = \frac{1}{3}x - 7$

e $y = \frac{3}{5}x + 4$, $y = -\frac{5}{3}x - 1$

f $y = \frac{5}{7}x$, $y = \frac{5}{7}x - 3$

g $y = 5x - 3$, $5x - y + 4 = 0$

h $5x - y - 1 = 0$, $y = -\frac{1}{5}x$

i $y = -\frac{3}{2}x + 8$, $2x - 3y - 9 = 0$

j $4x - 5y + 1 = 0$, $8x - 10y - 2 = 0$

k $3x + 2y - 12 = 0$, $2x + 3y - 6 = 0$

l $5x - y + 2 = 0$, $2x + 10y - 4 = 0$

2 Find an equation of the line that passes through the point $(6, -2)$ and is perpendicular to the line $y = 3x + 5$.

3 Find an equation of the line that passes through the point $(-2, 7)$ and is parallel to the line $y = 4x + 1$. Write your answer in the form $ax + by + c = 0$.

4 Find an equation of the line:

a parallel to the line $y = -2x - 5$, passing through $(-\frac{1}{2}, \frac{3}{2})$

b parallel to the line $x - 2y - 1 = 0$, passing through $(0, 0)$

c perpendicular to the line $y = x - 4$, passing through $(-1, -2)$

d perpendicular to the line $2x + y - 9 = 0$, passing through $(4, -6)$.

5 Find an equation of the line:

a parallel to the line $y = 3x + 6$, passing through $(-2, 5)$

b perpendicular to the line $y = 3x + 6$, passing through $(-2, 5)$

c parallel to the line $4x - 6y + 7 = 0$, passing through $(3, 4)$

d perpendicular to the line $4x - 6y + 7 = 0$, passing through $(3, 4)$.

6 Find an equation of the line that passes through the point $(5, -5)$ and is perpendicular to the line $y = \frac{2}{3}x + 5$. Write your answer in the form $ax + by + c = 0$, where a, b and c are integers.

7 Find an equation of the line that passes through the point $(-2, -3)$ and is perpendicular to the line $y = -\frac{4}{7}x + 5$. Write your answer in the form $ax + by + c = 0$, where a, b and c are integers.

8 The line r passes through the points $(1, 4)$ and $(6, 8)$ and the line s passes through the points $(5, -3)$ and $(20, 9)$. Show that the lines r and s are parallel.

9 The line l passes through the points $(-3, 0)$ and $(3, -2)$ and the line n passes through the points $(1, 8)$ and $(-1, 2)$. Show that the lines l and n are perpendicular.

10 The vertices of a quadrilateral $ABCD$ has coordinates $A(-1, 5)$, $B(7, 1)$, $C(5, -3)$, $D(-3, 1)$. Show that the quadrilateral is a rectangle.

Mixed exercise 5F

1 The points A and B have coordinates $(-4, 6)$ and $(2, 8)$ respectively. A line p is drawn through B perpendicular to AB to meet the y-axis at the point C.

a Find an equation of the line p.

b Determine the coordinates of C.

E

2 The line l has equation $2x - y - 1 = 0$.
The line m passes through the point $A(0, 4)$ and is perpendicular to the line l.

a Find an equation of m and show that the lines l and m intersect at the point $P(2, 3)$.

The line n passes through the point $B(3, 0)$ and is parallel to the line m.

b Find an equation of n and hence find the coordinates of the point Q where the lines l and n intersect. **E**

3 The line L_1 has gradient $\frac{1}{7}$ and passes through the point $A(2, 2)$. The line L_2 has gradient -1 and passes through the point $B(4, 8)$. The lines L_1 and L_2 intersect at the point C.

a Find an equation for L_1 and an equation for L_2.

b Determine the coordinates of C. **E**

4 The straight line passing through the point $P(2, 1)$ and the point $Q(k, 11)$ has gradient $-\frac{5}{12}$.

a Find the equation of the line in terms of x and y only.

b Determine the value of k. **E**

5 **a** Find an equation of the line l which passes through the points $A(1, 0)$ and $B(5, 6)$.

The line m with equation $2x + 3y = 15$ meets l at the point C.

b Determine the coordinates of the point C. **E**

6 The line L passes through the points $A(1, 3)$ and $B(-19, -19)$.
Find an equation of L in the form $ax + by + c = 0$, where a, b and c are integers. **E**

7 The straight line l_1 passes through the points A and B with coordinates $(2, 2)$ and $(6, 0)$ respectively.

a Find an equation of l_1.

The straight line l_2 passes through the point C with coordinates $(-9, 0)$ and has gradient $\frac{1}{4}$.

b Find an equation of l_2. **E**

8 The straight line l_1 passes through the points A and B with coordinates $(0, -2)$ and $(6, 7)$ respectively.

a Find the equation of l_1 in the form $y = mx + c$.

The straight line l_2 with equation $x + y = 8$ cuts the y-axis at the point C. The lines l_1 and l_2 intersect at the point D.

b Calculate the coordinates of the point D.

c Calculate the area of $\triangle ACD$. **E**

9 The points A and B have coordinates $(2, 16)$ and $(12, -4)$ respectively. A straight line l_1 passes through A and B.

a Find an equation for l_1 in the form $ax + by = c$.

The line l_2 passes through the point C with coordinates $(-1, 1)$ and has gradient $\frac{1}{3}$.

b Find an equation for l_2. **E**

10 The points $A(-1, -2)$, $B(7, 2)$ and $C(k, 4)$, where k is a constant, are the vertices of $\triangle ABC$. Angle ABC is a right angle.

a Find the gradient of AB.

b Calculate the value of k.

c Find an equation of the straight line passing through B and C. Give your answer in the form $ax + by + c = 0$, where a, b and c are integers. **E**

11 The straight line l passes through $A(1, 3\sqrt{3})$ and $B(2 + \sqrt{3}, 3 + 4\sqrt{3})$.

a Calculate the gradient of l giving your answer as a surd in its simplest form.

b Give the equation of l in the form $y = mx + c$, where constants m and c are surds given in their simplest form.

c Show that l meets the x-axis at the point $C(-2, 0)$. **E**

12 **a** Find an equation of the straight line passing through the points with coordinates $(-1, 5)$ and $(4, -2)$, giving your answer in the form $ax + by + c = 0$, where a, b and c are integers.

The line crosses the x-axis at the point A and the y-axis at the point B, and O is the origin.

b Find the area of $\triangle OAB$. **E**

13 The points A and B have coordinates $(k, 1)$ and $(8, 2k - 1)$ respectively, where k is a constant. Given that the gradient of AB is $\frac{1}{3}$:

a show that $k = 2$

b find an equation for the line through A and B. **E**

14 The straight line l_1 has equation $4y + x = 0$.
The straight line l_2 has equation $y = 2x - 3$.

a On the same axes, sketch the graphs of l_1 and l_2. Show clearly the coordinates of all points at which the graphs meet the coordinate axes.

The lines l_1 and l_2 intersect at the point A.

b Calculate, as exact fractions, the coordinates of A.

c Find an equation of the line through A which is perpendicular to l_1. Give your answer in the form $ax + by + c = 0$, where a, b and c are integers. **E**

15 The points A and B have coordinates $(4, 6)$ and $(12, 2)$ respectively.
The straight line l_1 passes through A and B.

a Find an equation for l_1 in the form $ax + by + c = 0$, where a, b and c are integers.

The straight line l_2 passes through the origin and has gradient -4.

b Write down an equation for l_2.

The lines l_1 and l_2 intersect at the point C.

c Find the coordinates of C. **E**

Summary of key points

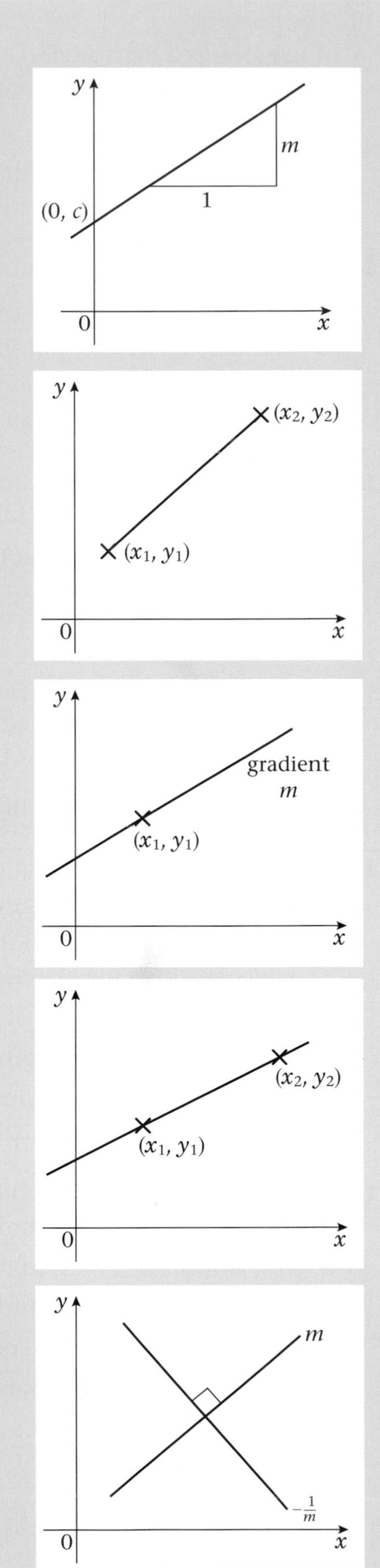

1 • In the general form

$$y = mx + c,$$

where m is the gradient and $(0, c)$ is the intercept on the y-axis.

• In the general form

$$ax + by + c = 0,$$

where a, b and c are integers.

2 You can work out the gradient m of the line joining the point with coordinates (x_1, y_1) to the point with coordinates (x_2, y_2) by using the formula

$$m = \frac{y_2 - y_1}{x_2 - x_1}$$

3 You can find the equation of a line with gradient m that passes through the point with coordinates (x_1, y_1) by using the formula

$$y - y_1 = m(x - x_1)$$

4 You can find the equation of the line that passes through the points with coordinates (x_1, y_1) and (x_2, y_2) by using the formula

$$\frac{y - y_1}{y_2 - y_1} = \frac{x - x_1}{x_2 - x_1}$$

5 If a line has a gradient m, a line perpendicular to it has a gradient of $-\frac{1}{m}$.

6 If two lines are perpendicular, the product of their gradients is -1.

After completing this chapter you should be able to

1. generate a sequence from the nth term, or from a recurrence relationship
2. know how to find the nth term of an arithmetic sequence, U_n
3. know how to find the sum to n terms of an arithmetic series, S_n
4. solve problems on arithmetic series using the formulae for U_n and S_n
5. know the meaning of the symbol $\sum$.

Sequences and series

Did you know?

...the famous story about a young boy named Carl Friedrich Gauss? His primary school teacher, J.G. Büttner, tried to occupy his pupils by making them add up the integers from 1 to 100. The young Gauss produced the correct answer within seconds. You will find out how he did it in this chapter.

Carl Friedrich Gauss (1777–1855), painted by Christian Albrecht Jensen

6.1 A series of numbers following a set rule is called a sequence. 3, 7, 11, 15, 19, … is an example of a sequence.

■ **Each number in a sequence is called a term.**

Example 1

Work out:

i the next three terms in each of the following sequences and **ii** the rule to find the next term.

a 14, 11, 8, 5, …

b 1, 2, 4, 8, …

c 1, 3, 7, 15, …

Look for the rule that takes you from one term to the next.

a 14, 11, 8, 5, …

To go from one term to the next you subtract 3.

i The next three terms are 2, −1 and −4.

ii

Term no.	1	2	3	4	5
Term	14	11	8	5	2

The rule to find the next term is 'subtract 3 from the previous term'.

b 1, 2, 4, 8, …

To go from one term to the next you multiply by 2.

i The next three terms are 16, 32 and 64.

ii

Term no.	1	2	3	4	5
Term	1	2	4	8	16

The rule to find the next term is 'multiply the previous term by 2'.

c 1, 3, 7, 15, …

To go from one term to the next you multiply by 2, then add 1.

i The next three terms are 31, 63 and 127.

ii

Term no.	1	2	3	4	5
Term	1	3	7	15	31

The rule to find the next term is 'multiply the previous term by 2 then add 1'.

Exercise 6A

Work out the next three terms of the following sequences. State the rule to find the next term in each case:

1 4, 9, 14, 19, ...

2 2, −2, 2, −2, ...

3 30, 27, 24, 21, ...

4 2, 6, 18, 54, ...

5 4, −2, 1, $-\frac{1}{2}$, ...

6 1, 2, 5, 14, ...

7 1, 1, 2, 3, 5, ...

8 1, $\frac{2}{3}$, $\frac{3}{5}$, $\frac{4}{7}$, ...

9 4, 3, 2.5, 2.25, 2.125, ...

10 0, 3, 8, 15, ...

Hints: Question 6 – Look for two operations. Question 8 – Treat numerator and denominator separately.

6.2 When you know a formula for the *n*th term of a sequence (e.g. $U_n = 3n - 1$) you can use this to find any term in the sequence.

■ **The *n*th term of a sequence is sometimes called the general term.**

Example 2

The nth term of a sequence is given by $U_n = 3n - 1$.
Work out:

a The first term. **b** The third term. **c** The nineteenth term.

a $U_1 = 3 \times 1 - 1$
$= 2$

Substitute $n = 1$

b $U_3 = 3 \times 3 - 1$
$= 8$

Substitute $n = 3$

c $U_{19} = 3 \times 19 - 1$
$= 56$

Substitute $n = 19$

Example 3

The nth term of a sequence is given by $U_n = \frac{n^2}{(n+1)}$.

Work out:

a The first three terms. **b** The 49th term.

a $U_1 = \frac{1 \times 1}{1 + 1} = \frac{1}{2}$ (Substitute $n = 1$)

$U_2 = \frac{2 \times 2}{2 + 1} = \frac{4}{3}$ (Substitute $n = 2$)

$U_3 = \frac{3 \times 3}{3 + 1} = \frac{9}{4}$ (Substitute $n = 3$)

Use $U_n = \frac{n^2}{n+1}$ with $n = 1$, 2 and 3.

b $U_{49} = \frac{49 \times 49}{49 + 1}$ (Substitute $n = 49$)

$= \frac{2401}{50}$

Use $U_n = \frac{n^2}{n+1}$ with $n = 49$.

Example 4

Find the value of n for which U_n has the given value:

a $U_n = 5n - 2$, $U_n = 153$

b $U_n = n^2 + 5$, $U_n = 149$

c $U_n = n^2 - 7n + 12$, $U_n = 72$

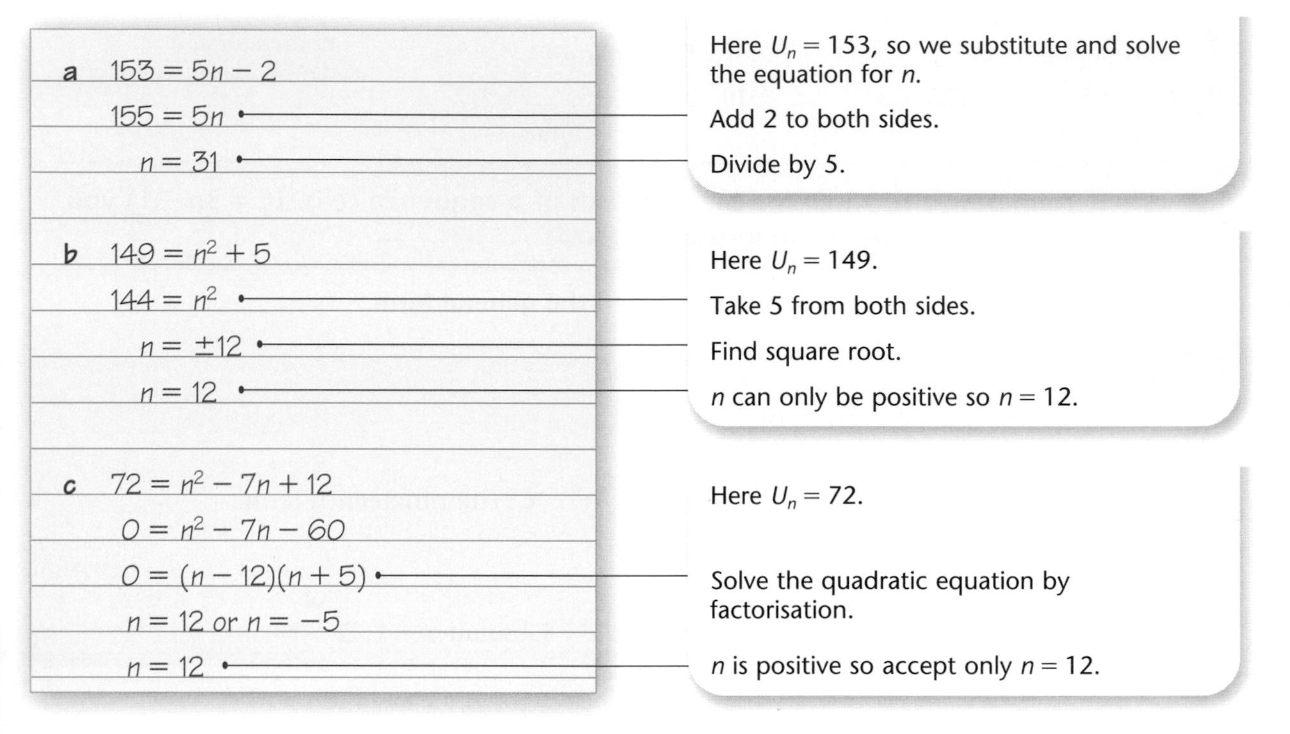

Example 5

A sequence is generated by the formula $U_n = an + b$ where a and b are constants to be found. Given that $U_3 = 5$ and $U_8 = 20$, find the values of the constants a and b.

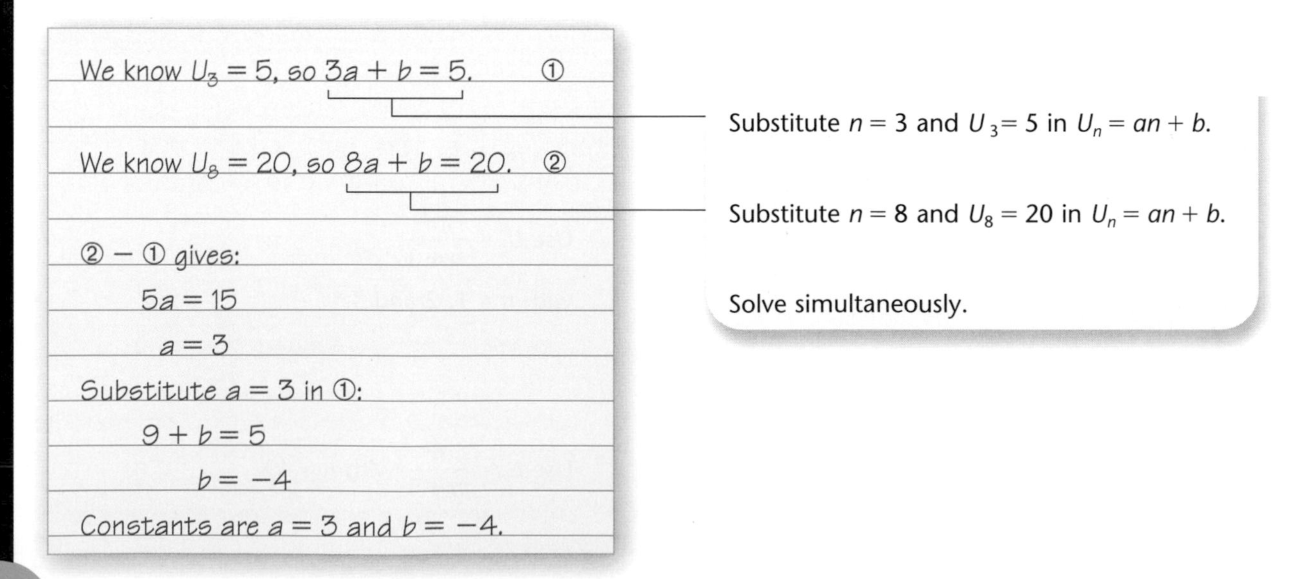

Exercise 6B

1 Find the U_1, U_2, U_3 and U_{10} of the following sequences, where:

a $U_n = 3n + 2$

b $U_n = 10 - 3n$

c $U_n = n^2 + 5$

d $U_n = (n - 3)^2$

e $U_n = (-2)^n$

f $U_n = \frac{n}{n+2}$

g $U_n = (-1)^n \frac{n}{n+2}$

h $U_n = (n - 2)^3$

2 Find the value of n for which U_n has the given value:

a $U_n = 2n - 4$, $U_n = 24$

b $U_n = (n - 4)^2$, $U_n = 25$

c $U_n = n^2 - 9$, $U_n = 112$

d $U_n = \frac{2n+1}{n-3}$, $U_n = \frac{19}{6}$

e $U_n = n^2 + 5n - 6$, $U_n = 60$

f $U_n = n^2 - 4n + 11$, $U_n = 56$

g $U_n = n^2 + 4n - 5$, $U_n = 91$

h $U_n = (-1)^n \frac{n}{n+4}$, $U_n = \frac{7}{9}$

i $U_n = \frac{n^3+3}{5}$, $U_n = 13.4$

j $U_n = \frac{n^3}{5} + 3$, $U_n = 28$

3 Prove that the $(2n + 1)$th term of the sequence $U_n = n^2 - 1$ is a multiple of 4.

4 Prove that the terms of the sequence $U_n = n^2 - 10n + 27$ are all positive. For what value of n is U_n smallest?

Hint: Question 4 – Complete the square.

5 A sequence is generated according to the formula $U_n = an + b$, where a and b are constants. Given that $U_3 = 14$ and $U_5 = 38$, find the values of a and b.

6 A sequence is generated according to the formula $U_n = an^2 + bn + c$, where a, b and c are constants. If $U_1 = 4$, $U_2 = 10$ and $U_3 = 18$, find the values of a, b and c.

7 A sequence is generated from the formula $U_n = pn^3 + q$, where p and q are constants. Given that $U_1 = 6$ and $U_3 = 19$, find the values of the constants p and q.

6.3 When you know the rule to get from one term to the next, you can use this information to produce a recurrence relationship (or recurrence formula).

Look at the following sequence of numbers:

5, 8, 11, 14, 17, ...

We can describe this by the rule 'add 3 to the previous term'.
We can see that:

$U_2 = U_1 + 3$
$U_3 = U_2 + 3$
$U_4 = U_3 + 3$
etc.

This sequence can also be described by the recurrence formula:

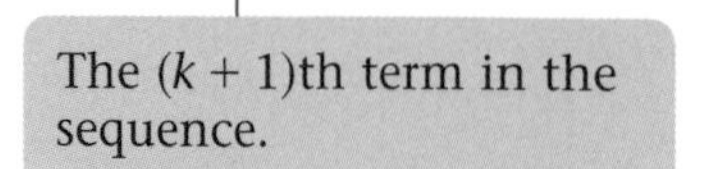

$$U_{k+1} = U_k + 3 \quad (k \geqslant 1)$$

It works for all values of k bigger than or equal to 1.

The $(k + 1)$th term in the sequence.

The kth term in the sequence.

You must always state the first term of the sequence, as many different sequences have the same recurrence relationship. For example, the sequences

4, 7, 10, 13, 16, ...

and

5, 8, 11, 14, 17, ...

could both be described by the recurrence formula $U_{k+1} = U_k + 3$, but we can distinguish between them by stating

$U_{k+1} = U_k + 3$, $k \geqslant 1$ with $U_1 = 4$ in the first example

but

$U_{k+1} = U_k + 3$, $k \geqslant 1$ and $U_1 = 5$ in the second example.

■ **A sequence can be expressed by a recurrence relationship. For example, the sequence 5, 9, 13, 17, ... can be formed from $U_{n+1} = U_n + 4$, $U_1 = 5$ (U_1 must be given).**

Example 6

Find the first four terms of the following sequences:

a $U_{n+1} = U_n + 4,\ U_1 = 7$ **b** $U_{n+1} = U_n + 4,\ U_1 = 5$ **c** $U_{n+2} = 3U_{n+1} - U_n,\ U_1 = 4$ and $U_2 = 2$

a $U_{n+1} = U_n + 4,\ U_1 = 7$

Substituting $n = 1$, $U_2 = U_1 + 4 = 7 + 4 = 11$.

Substituting $n = 2$, $U_3 = U_2 + 4 = 11 + 4 = 15$.

Substituting $n = 3$, $U_4 = U_3 + 4 = 15 + 4 = 19$.

Sequence is 7, 11, 15, 19, ...

Substitute $n = 1$, 2 and 3. As you are given U_1 you have the first term.

b $U_{n+1} = U_n + 4,\ U_1 = 5$

Substituting $n = 1$, $U_2 = U_1 + 4 = 5 + 4 = 9$.

Substituting $n = 2$, $U_3 = U_2 + 4 = 9 + 4 = 13$.

Substituting $n = 3$, $U_4 = U_3 + 4 = 13 + 4 = 17$.

Sequence is 5, 9, 13, 17, ...

This is the same recurrence formula. It produces a different sequence because U_1 is different.

c $U_{n+2} = 3U_{n+1} - U_n,\ U_1 = 4,\ U_2 = 2.$

Substituting $n = 1$, $U_3 = 3U_2 - U_1 = 3 \times 2 - 4 = 2$.

Substituting $n = 2$, $U_4 = 3U_3 - U_2 = 3 \times 2 - 2 = 4$.

Sequence is 4, 2, 2, 4, ...

This formula links up three terms. Simply substitute in the values of n to see how the relationship works.

Example 7

A sequence of terms $\{U_n\}$, $n \geqslant 1$ is defined by the recurrence relation $U_{n+2} = mU_{n+1} + U_n$ where m is a constant. Given also that $U_1 = 2$ and $U_2 = 5$:

a find an expression in terms of m for U_3

b find an expression in terms of m for U_4.

Given the value of $U_4 = 21$:

c find the possible values of m.

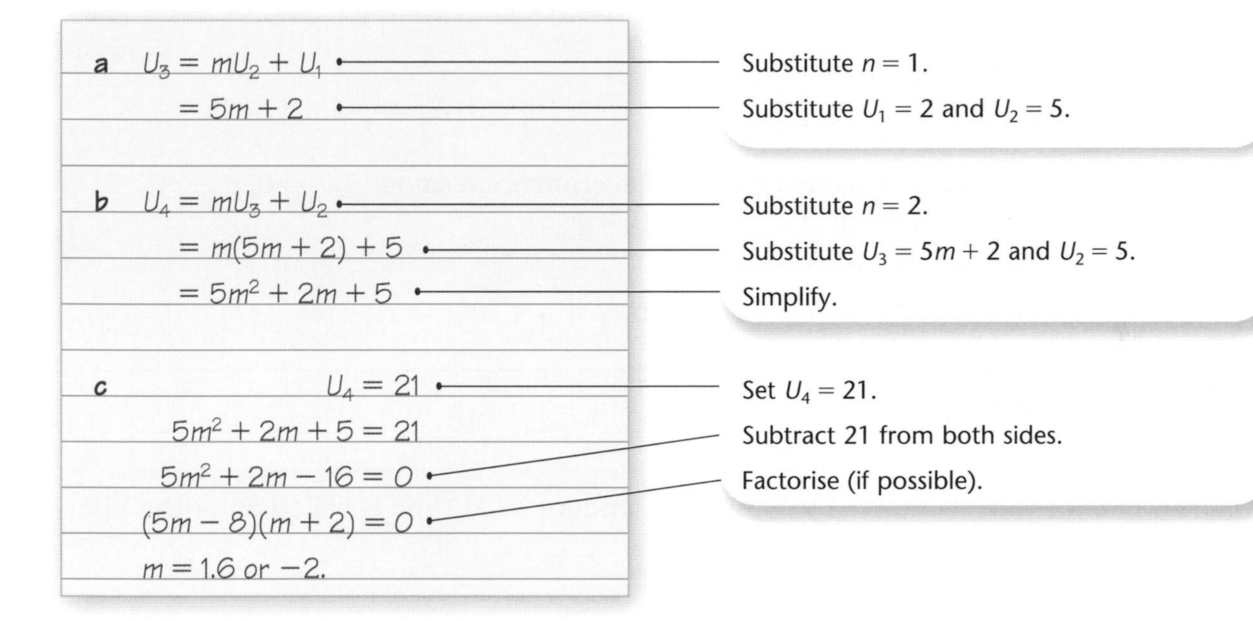

Exercise 6C

1 Find the first four terms of the following recurrence relationships:

a $U_{n+1} = U_n + 3$, $U_1 = 1$

b $U_{n+1} = U_n - 5$, $U_1 = 9$

c $U_{n+1} = 2U_n$, $U_1 = 3$

d $U_{n+1} = 2U_n + 1$, $U_1 = 2$

e $U_{n+1} = \frac{U_n}{2}$, $U_1 = 10$

f $U_{n+1} = (U_n)^2 - 1$, $U_1 = 2$

g $U_{n+2} = 2U_{n+1} + U_n$, $U_1 = 3$, $U_2 = 5$

2 Suggest possible recurrence relationships for the following sequences (remember to state the first term):

a 3, 5, 7, 9, …

b 20, 17, 14, 11, …

c 1, 2, 4, 8, …

d 100, 25, 6.25, 1.5625, …

e 1, −1, 1, −1, 1, …

f 3, 7, 15, 31, …

g 0, 1, 2, 5, 26, …

h 26, 14, 8, 5, 3.5, …

i 1, 1, 2, 3, 5, 8, 13, …

j 4, 10, 18, 38, 74, …

3 By writing down the first four terms or otherwise, find the recurrence formula that defines the following sequences:

a $U_n = 2n - 1$

b $U_n = 3n + 2$

c $U_n = n + 2$

d $U_n = \frac{n+1}{2}$

e $U_n = n^2$

f $U_n = (-1)^n n$

4 A sequence of terms $\{U_n\}$ is defined $n \geqslant 1$ by the recurrence relation $U_{n+1} = kU_n + 2$, where k is a constant. Given that $U_1 = 3$:

a find an expression in terms of k for U_2

b hence find an expression for U_3.

Given that $U_3 = 42$:

c find possible values of k.

5 A sequence of terms $\{U_k\}$ is defined $k \geqslant 1$ by the recurrence relation $U_{k+2} = U_{k+1} - pU_k$, where p is a constant. Given that $U_1 = 2$ and $U_2 = 4$:

a find an expression in terms of p for U_3

b hence find an expression in terms of p for U_4.

Given also that U_4 is twice the value of U_3:

c find the value of p.

6.4 A sequence that increases by a constant amount each time is called an arithmetic sequence.

The following are examples of arithmetic sequences:

3, 7, 11, 15, 19, ... (because you add 4 each time)
2, 7, 12, 17, 22, ... (because you add 5 each time)
17, 14, 11, 8, ... (because you add −3 each time)
$a, a + d, a + 2d, a + 3d, \ldots$ (because you add d each time)

■ **A recurrence relationship of the form**

$$U_{k+1} = U_k + n,\ k \geqslant 1\ \ n \in \mathbb{Z}$$

is called an arithmetic sequence.

Example 8

Find the **a** 10th, **b** nth and **c** 50th terms of the arithmetic sequence 3, 7, 11, 15, 19, ...

Sequence is 3, 7, 11, 15, ...
First term = 3
Second term = 3 + 4
Third term = 3 + 4 + 4
Fourth term = 3 + 4 + 4 + 4

The sequence is going up in fours.
It is starting at 3.
The first term is 3 + 0 × 4.
The second term is 3 + 1 × 4.
The third term is 3 + 2 lots of 4.
The fourth term is 3 + 3 lots of 4.

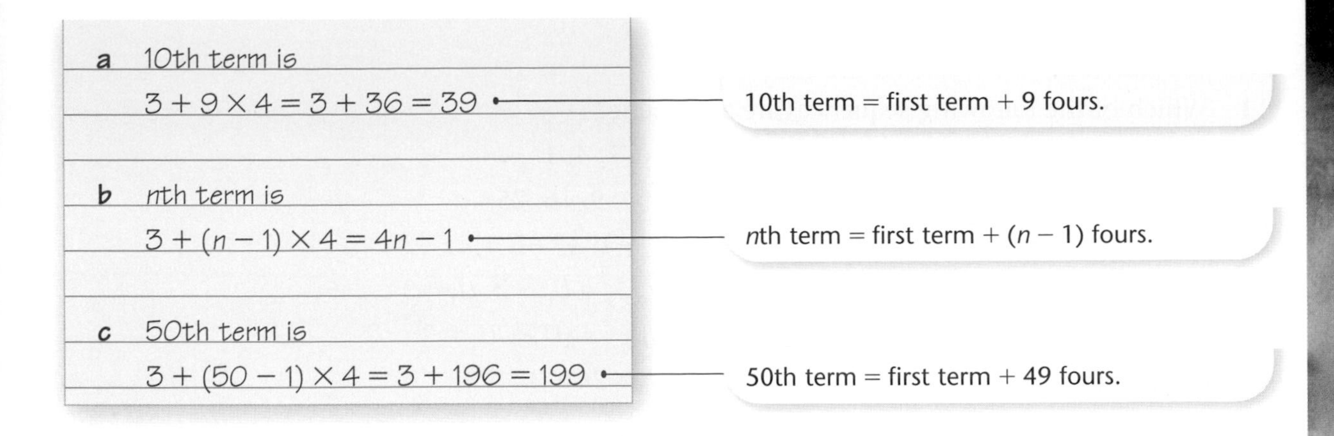

Example 9

A 6 metre high tree is planted in a garden. If it grows 1.5 metres a year:

a How high will it be after it has been in the garden for 8 years?

b After how many years will it be 24 metres high?

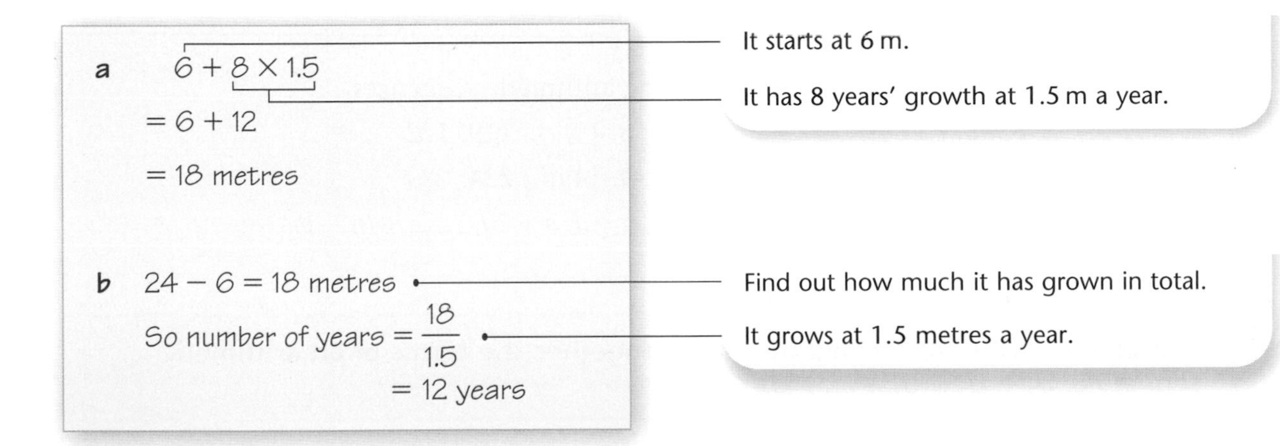

Example 10

Find the number of terms in the arithmetic sequence 7, 11, 15, …, 143:

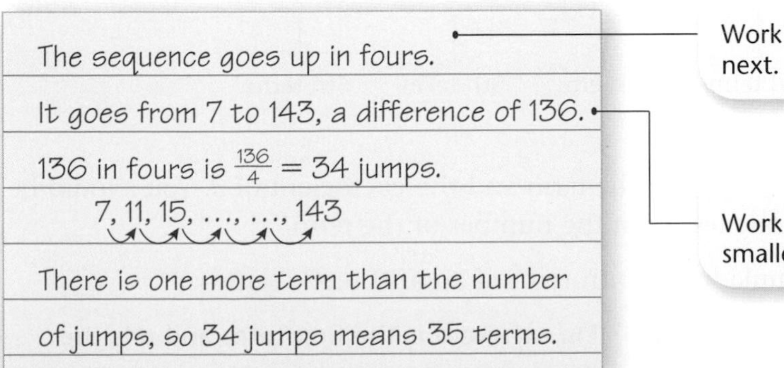

Work out how to get from one term to the next.

Work out the difference between largest and smallest numbers.

Exercise 6D

1 Which of the following sequences are arithmetic?

a 3, 5, 7, 9, 11, ...
b 10, 7, 4, 1, ...
c $y, 2y, 3y, 4y, \ldots$
d 1, 4, 9, 16, 25, ...
e 16, 8, 4, 2, 1, ...
f 1, −1, 1, −1, 1, ...
g $y, y^2, y^3, y^4, \ldots$
h $U_{n+1} = U_n + 2, U_1 = 3$
i $U_{n+1} = 3U_n - 2, U_1 = 4$
j $U_{n+1} = (U_n)^2, U_1 = 2$
k $U_n = n(n+1)$
l $U_n = 2n + 3$

2 Find the 10th and nth terms in the following arithmetic progressions:

a 5, 7, 9, 11, ...
b 5, 8, 11, 14, ...
c 24, 21, 18, 15, ...
d −1, 3, 7, 11, ...
e $x, 2x, 3x, 4x, \ldots$
f $a, a + d, a + 2d, a + 3d, \ldots$

3 An investor puts £4000 in an account. Every month thereafter she deposits another £200. How much money in total will she have invested at the start of **a** the 10th month and **b** the mth month? (Note that at the start of the 6th month she will have made only 5 deposits of £200.)

4 Calculate the number of terms in the following arithmetic sequences:

a 3, 7, 11, ..., 83, 87
b 5, 8, 11, ..., 119, 122
c 90, 88, 86, ..., 16, 14
d 4, 9, 14, ..., 224, 229
e $x, 3x, 5x, \ldots, 35x$
f $a, a + d, a + 2d, \ldots, a + (n-1)d$

6.5 Arithmetic series are formed by adding together the terms of an arithmetic sequence, $U_1 + U_2 + U_3 + \ldots + U_n$.

In an arithmetic series the next term is found by adding (or subtracting) a constant number. This number is called the common difference d.
The first term is represented by a.

■ **Therefore all arithmetic series can be put in the form**

$$a + (a + d) + (a + 2d) + (a + 3d) + (a + 4d) + (a + 5d)$$

1st term, 2nd term, 3rd term, 4th term, 5th term, 6th term

Look at the relationship between the number of the term and the coefficient of d. You should be able to see that the coefficient of d is one less than the number of the term.

We can use this fact to produce a formula for the nth term of an arithmetic series.

■ **The nth term of an arithmetic series is $a + (n - 1)d$, where a is the first term and d is the common difference.**

Example 11

Find **i** the 20th and **ii** the 50th terms of the following series:

a $4 + 7 + 10 + 13 + \ldots$ **b** $100 + 93 + 86 + 79 + \ldots$

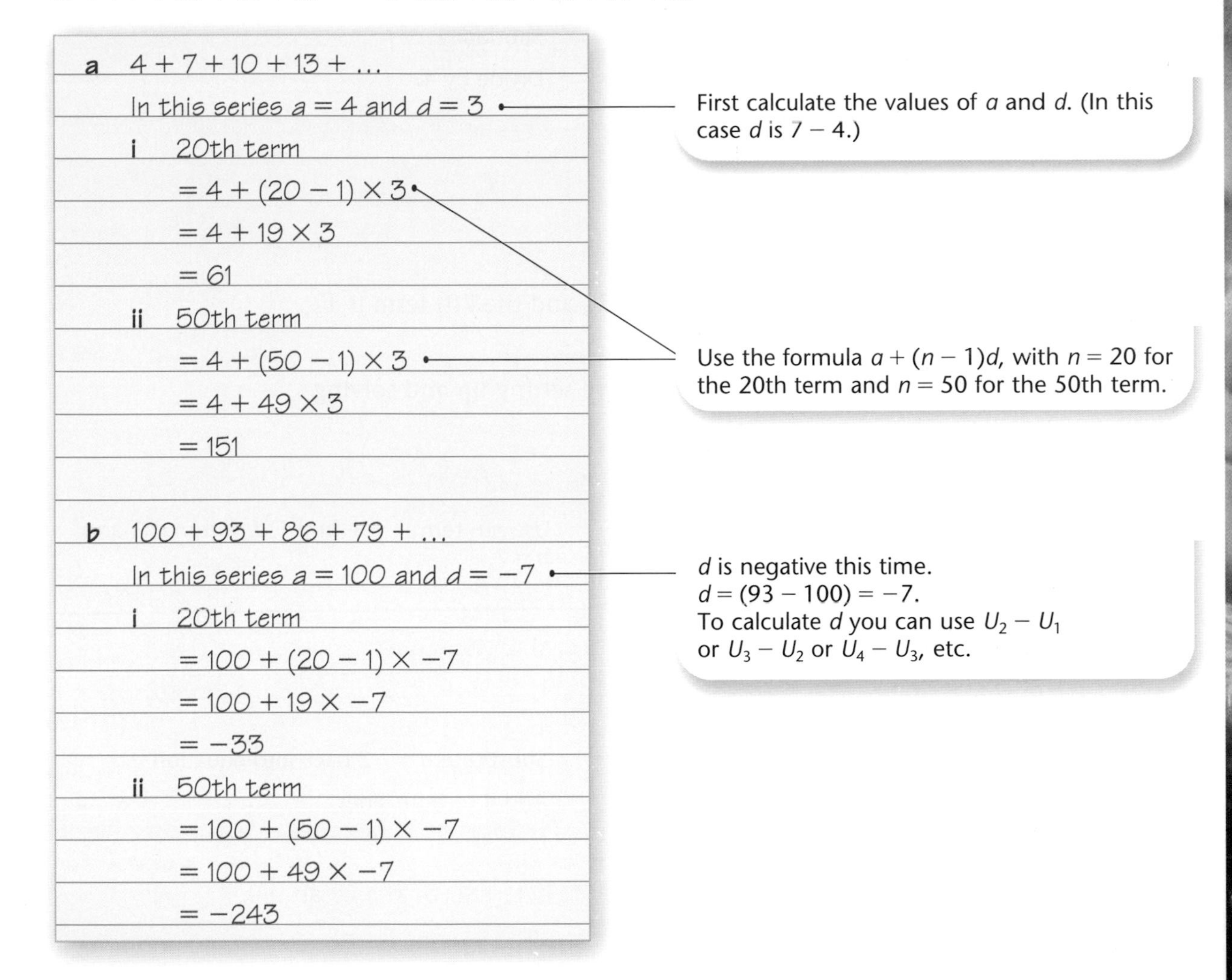

a $4 + 7 + 10 + 13 + \ldots$

In this series $a = 4$ and $d = 3$

i 20th term

$= 4 + (20 - 1) \times 3$

$= 4 + 19 \times 3$

$= 61$

ii 50th term

$= 4 + (50 - 1) \times 3$

$= 4 + 49 \times 3$

$= 151$

b $100 + 93 + 86 + 79 + \ldots$

In this series $a = 100$ and $d = -7$

i 20th term

$= 100 + (20 - 1) \times -7$

$= 100 + 19 \times -7$

$= -33$

ii 50th term

$= 100 + (50 - 1) \times -7$

$= 100 + 49 \times -7$

$= -243$

First calculate the values of a and d. (In this case d is $7 - 4$.)

Use the formula $a + (n - 1)d$, with $n = 20$ for the 20th term and $n = 50$ for the 50th term.

d is negative this time.
$d = (93 - 100) = -7$.
To calculate d you can use $U_2 - U_1$ or $U_3 - U_2$ or $U_4 - U_3$, etc.

Example 12

For the arithmetic series $5 + 9 + 13 + 17 + 21 + \ldots + 805$:

a find the number of terms **b** which term of the series would be 129?

Series is $5 + 9 + 13 + 17 + 21 + \ldots + 805$.

In this series $a = 5$ and $d = 4$.

a Using nth term $= a + (n - 1)d$

$805 = 5 + (n - 1) \times 4$

$805 = 5 + 4n - 4$

$805 = 4n + 1$

$804 = 4n$

$n = 201$

There are 201 terms in this series.

A good starting point in all questions is to find the values of a and d.
Here $a = 5$ and $a + d = 9$, so $d = 4$.

The nth term is $a + (n - 1)d$.
So replace U_n with 805 and solve for n.

Subtract 1.

Divide by 4.

b Using nth term $= a + (n-1)d$

$129 = 5 + (n-1) \times 4$

$129 = 4n + 1$

$128 = 4n$ — Subtract 1.

$n = 32$ — Divide by 4.

The 32nd term is 129.

This time the nth term is 129. So replace U_n with 129.

Example 13

Given that the 3rd term of an arithmetic series is 20 and the 7th term is 12:

a find the first term **b** find the 20th term.

(Note: These are very popular questions and involve setting up and solving simultaneous equations.)

a 3rd term = 20, so $a + 2d = 20$. ①

7th term = 12, so $a + 6d = 12$. ②

Use nth term $= a + (n-1)d$, with $n = 3$ and $n = 7$.

Taking ① from ②:

$4d = -8$

$d = -2$

The common difference is -2.

Substitute $d = -2$ back into equation ①.

$a + 2 \times -2 = 20$

$a - 4 = 20$

Add 4 to both sides.

$a = 24$

The first term is 24.

b 20th term $= a + 19d$ — Use nth term is $a + (n-1)d$ with $n = 20$.

$= 24 + 19 \times -2$ — Substitute a = 24 and $d = -2$.

$= 24 - 38$

$= -14$

The 20th term is -14.

Exercise 6E

1 Find **i** the 20th and **ii** the nth terms of the following arithmetic series:

a $2 + 6 + 10 + 14 + 18 \ldots$

b $4 + 6 + 8 + 10 + 12 + \ldots$

c $80 + 77 + 74 + 71 + \ldots$

d $1 + 3 + 5 + 7 + 9 + \ldots$

e $30 + 27 + 24 + 21 + \ldots$

f $2 + 5 + 8 + 11 + \ldots$

g $p + 3p + 5p + 7p + \ldots$

h $5x + x + (-3x) + (-7x) + \ldots$

2 Find the number of terms in the following arithmetic series:

a $5 + 9 + 13 + 17 + \ldots + 121$

b $1 + 1.25 + 1.5 + 1.75 \ldots + 8$

c $-4 + -1 + 2 + 5 \ldots + 89$

d $70 + 61 + 52 + 43 \ldots + -200$

e $100 + 95 + 90 + \ldots + (-1000)$

f $x + 3x + 5x \ldots + 153x$

3 The first term of an arithmetic series is 14. If the fourth term is 32, find the common difference.

4 Given that the 3rd term of an arithmetic series is 30 and the 10th term is 9 find a and d. Hence find which term is the first one to become negative.

5 In an arithmetic series the 20th term is 14 and the 40th term is -6. Find the 10th term.

6 The first three terms of an arithmetic series are $5x$, 20 and $3x$. Find the value of x and hence the values of the three terms.

7 For which values of x would the expression -8, x^2 and $17x$ form the first three terms of an arithmetic series?

Hint: Question 6 – Find two expressions equal to the common difference and set them equal to each other.

6.6 You need to be able to find the sum of an arithmetic series.

The method of finding this sum is attributed to a famous mathematician called Carl Friedrich Gauss (1777–1855). He reputedly solved the following sum whilst in Junior School:

$$1 + 2 + 3 + 4 + 5 + \ldots + 99 + 100$$

Here is how he was able to work it out:

Let $S = 1 + 2 + 3 + 4 \ldots + 98 + 99 + 100$

Reversing the sum $S = 100 + 99 + 98 + 97 \ldots + 3 + 2 + 1$

Adding the two sums $2S = 101 + 101 + 101 + \ldots + 101 + 101 + 101$

$$2S = 100 \times 101$$
$$S = (100 \times 101) \div 2$$
$$S = 5050$$

In general:

$$S_n = a + (a + d) + (a + 2d) + \ldots + (a + (n - 2)d) + (a + (n - 1)d)$$

Reversing the sum:

$$S_n = (a + (n - 1)d) + (a + (n - 2)d) + (a + (n - 3)d) + \ldots + (a + d) + a$$

Adding the two sums:

$$2S_n = [2a + (n - 1)d] + [2a + (n - 1)d] + \ldots + [2a + (n - 1)d]$$
$$2S_n = n[2a + (n - 1)d]$$
$$S_n = \frac{n}{2}[2a + (n - 1)d]$$

Hint: There are n lots of $2a + (n - 1)d$.

Prove for yourself that it could be $S_n = \frac{n}{2}(a + L)$ where $L = a + (n - 1)d$.

■ **The formula for the sum of an arithmetic series is**

$$S_n = \frac{n}{2}[2a + (n - 1)d]$$

or $$S_n = \frac{n}{2}(a + L)$$

where a is the first term, d is the common difference, n is the number of terms and L is the last term in the series.

You could be asked to prove these formulae.

Example 14

Find the sum of the first 100 odd numbers.

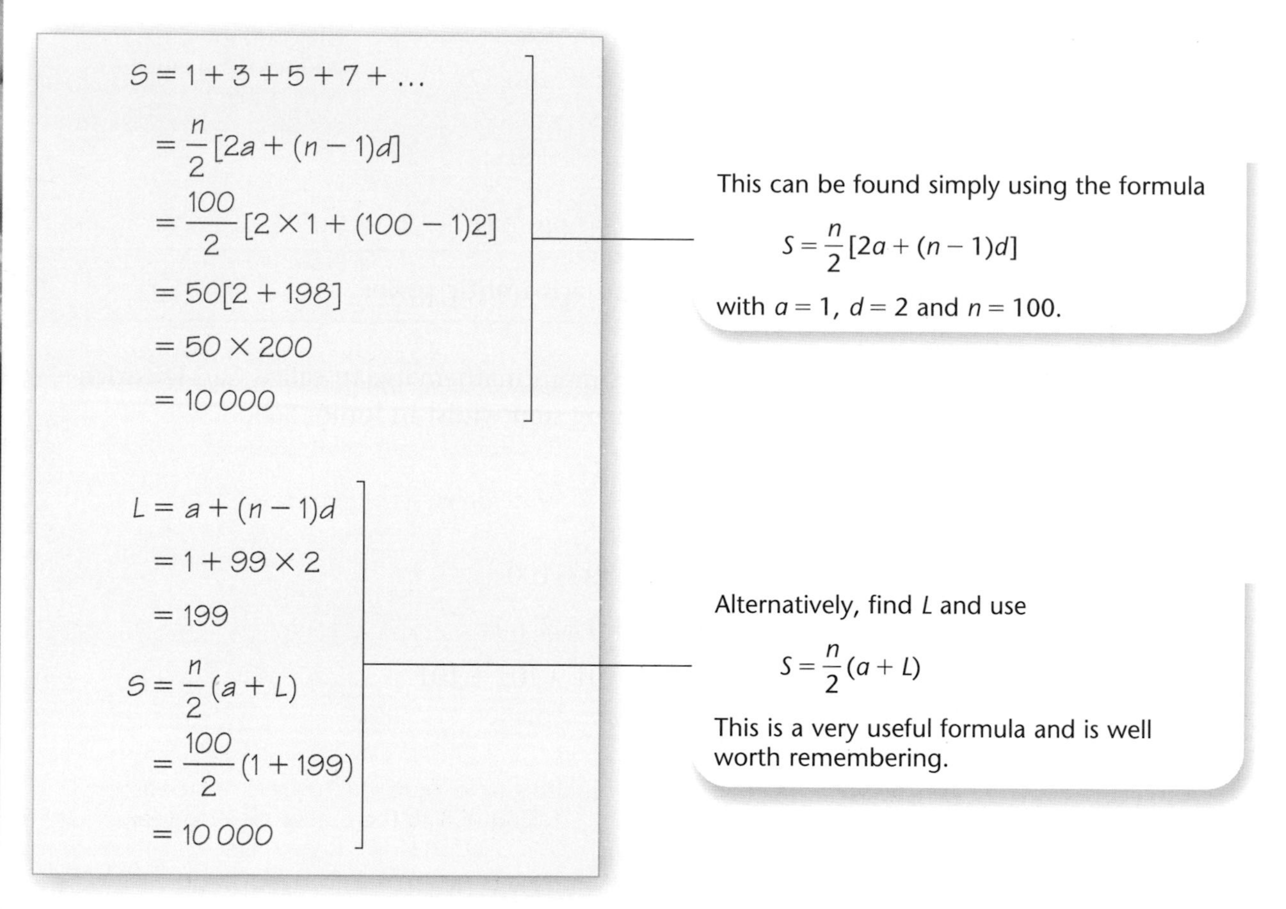

$$S = 1 + 3 + 5 + 7 + \ldots$$
$$= \frac{n}{2}[2a + (n - 1)d]$$
$$= \frac{100}{2}[2 \times 1 + (100 - 1)2]$$
$$= 50[2 + 198]$$
$$= 50 \times 200$$
$$= 10\,000$$

This can be found simply using the formula

$$S = \frac{n}{2}[2a + (n - 1)d]$$

with $a = 1$, $d = 2$ and $n = 100$.

$$L = a + (n - 1)d$$
$$= 1 + 99 \times 2$$
$$= 199$$
$$S = \frac{n}{2}(a + L)$$
$$= \frac{100}{2}(1 + 199)$$
$$= 10\,000$$

Alternatively, find L and use

$$S = \frac{n}{2}(a + L)$$

This is a very useful formula and is well worth remembering.

Example 15

Find the least number of terms required for the sum of $4 + 9 + 14 + 19 + \ldots$ to exceed 2000.

Always establish what you are given in a question. As you are adding on positive terms, it is easier to solve the equality $S_n = 2000$.

$4 + 9 + 14 + 19 + \ldots > 2000$

Using $S = \frac{n}{2}[2a + (n - 1)d]$

$2000 = \frac{n}{2}[2 \times 4 + (n - 1)5]$

$4000 = n(8 + 5n - 5)$

$4000 = n(5n + 3)$

$4000 = 5n^2 + 3n$

$0 = 5n^2 + 3n - 4000$

$n = \frac{-3 \pm \sqrt{(9 + 80\,000)}}{10}$

$= 27.9, -28.5$

28 terms are needed.

Knowing $a = 4$, $d = 5$ and $S_n = 2000$, you need to find n.

Substitute into $S = \frac{n}{2}[2a + (n - 1)d]$.

Solve using formula $n = \frac{-b \pm \sqrt{(b^2 - 4ac)}}{2a}$.

Accept positive answer and round up.

Example 16

Robert starts his new job on a salary of £15 000. He is promised rises of £1000 a year, at the end of every year, until he reaches his maximum salary of £25 000. Find his total earnings (since appointed) after **a** 8 years with the firm and **b** 14 years with the firm.

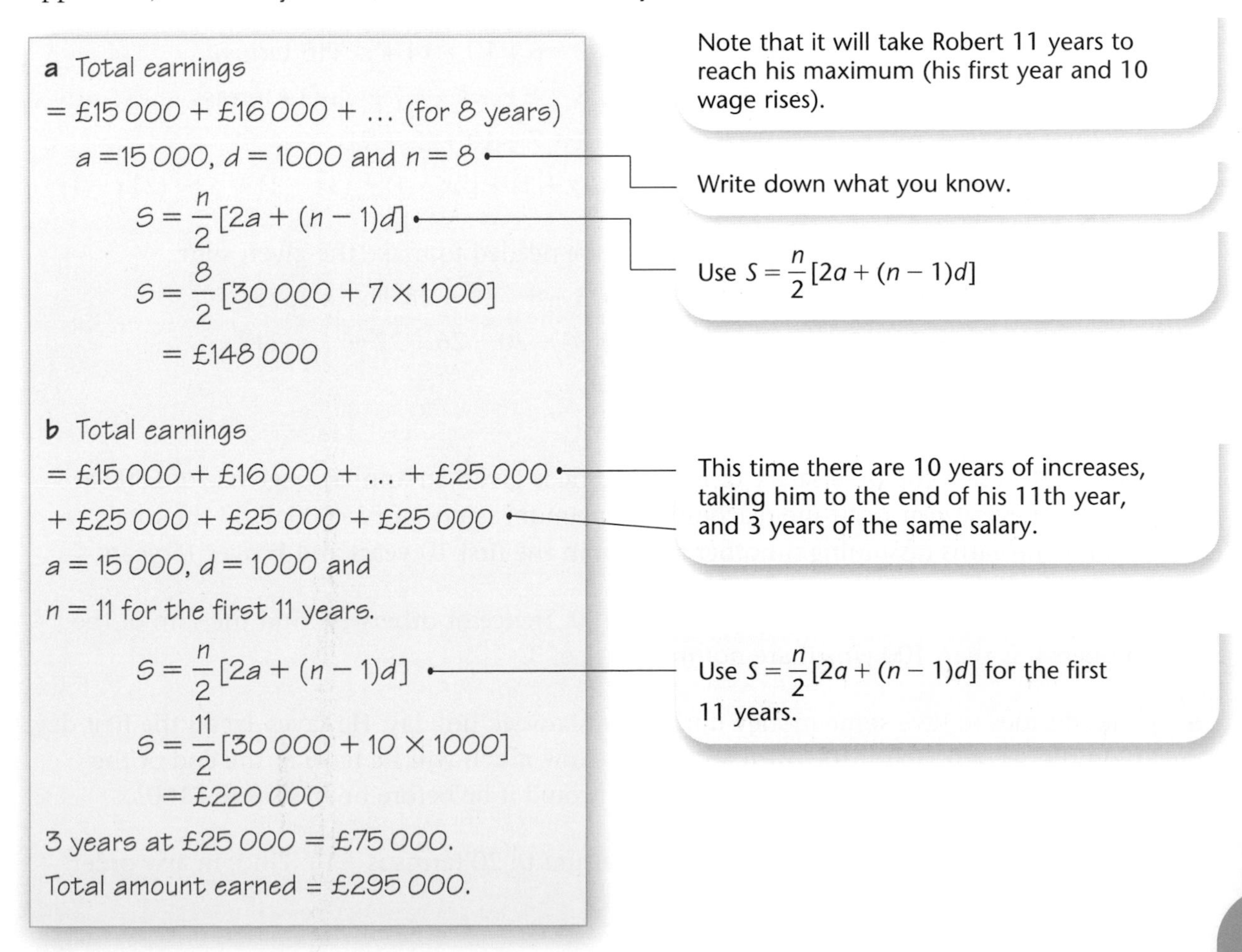

a Total earnings

$= £15\,000 + £16\,000 + \ldots$ (for 8 years)

$a = 15\,000$, $d = 1000$ and $n = 8$

$S = \frac{n}{2}[2a + (n - 1)d]$

$S = \frac{8}{2}[30\,000 + 7 \times 1000]$

$= £148\,000$

b Total earnings

$= £15\,000 + £16\,000 + \ldots + £25\,000$

$+ £25\,000 + £25\,000 + £25\,000$

$a = 15\,000$, $d = 1000$ and

$n = 11$ for the first 11 years.

$S = \frac{n}{2}[2a + (n - 1)d]$

$S = \frac{11}{2}[30\,000 + 10 \times 1000]$

$= £220\,000$

3 years at £25 000 = £75 000.

Total amount earned = £295 000.

Note that it will take Robert 11 years to reach his maximum (his first year and 10 wage rises).

Write down what you know.

Use $S = \frac{n}{2}[2a + (n - 1)d]$

This time there are 10 years of increases, taking him to the end of his 11th year, and 3 years of the same salary.

Use $S = \frac{n}{2}[2a + (n - 1)d]$ for the first 11 years.

Example 17

Show that the sum of the first n natural numbers is $\frac{1}{2}n(n+1)$.

$S = 1 + 2 + 3 + 4 + \ldots + n$

This is an arithmetic series with

$a = 1, d = 1, n = n.$

$$S = \frac{n}{2}[2a + (n-1)d]$$

$$S = \frac{n}{2}[2 \times 1 + (n-1) \times 1]$$

$$S = \frac{n}{2}(2 + n - 1)$$

$$S = \frac{n}{2}(n+1)$$

$$= \frac{1}{2}n(n+1)$$

Use $S = \frac{n}{2}[2a + (n-1)d]$ with $a = 1$, $d = 1$ and $n = n$.

Exercise 6F

1 Find the sums of the following series:

a $3 + 7 + 11 + 15 + \ldots$ (20 terms)

b $2 + 6 + 10 + 14 + \ldots$ (15 terms)

c $30 + 27 + 24 + 21 + \ldots$ (40 terms)

d $5 + 1 + -3 + -7 + \ldots$ (14 terms)

e $5 + 7 + 9 + \ldots + 75$

f $4 + 7 + 10 + \ldots + 91$

g $34 + 29 + 24 + 19 + \ldots + -111$

h $(x + 1) + (2x + 1) + (3x + 1) + \ldots + (21x + 1)$

2 Find how many terms of the following series are needed to make the given sum:

a $5 + 8 + 11 + 14 + \ldots = 670$

b $3 + 8 + 13 + 18 + \ldots = 1575$

c $64 + 62 + 60 + \ldots = 0$

d $34 + 30 + 26 + 22 + \ldots = 112$

3 Find the sum of the first 50 even numbers.

4 Carol starts a new job on a salary of £20 000. She is given an annual wage rise of £500 at the end of every year until she reaches her maximum salary of £25 000. Find the total amount she earns (assuming no other rises), **a** in the first 10 years and **b** over 15 years.

5 Find the sum of the multiples of 3 less than 100. Hence or otherwise find the sum of the numbers less than 100 which are not multiples of 3.

6 James decides to save some money during the six-week holiday. He saves 1p on the first day, 2p on the second, 3p on the third and so on. How much will he have at the end of the holiday (42 days)? If he carried on, how long would it be before he has saved £100?

7 The first term of an arithmetic series is 4. The sum to 20 terms is -15. Find, in any order, the common difference and the 20th term.

8 The sum of the first three numbers of an arithmetic series is 12. If the 20th term is -32, find the first term and the common difference.

9 Show that the sum of the first $2n$ natural numbers is $n(2n + 1)$.

10 Prove that the sum of the first n odd numbers is n^2.

6.7 You can use $\sum$ to signify 'the sum of'.

For example:

$\sum_{n=1}^{10} 2n$ means sum of $2n$ from $n = 1$ to $n = 10$

$= 2 + 4 + 6 + 8 + 10 + 12 + 14 + 16 + 18 + 20$

$\sum_{n=1}^{10} U_n = U_1 + U_2 + U_3 + \ldots + U_{10}$

$\sum_{r=0}^{10} (2 + 3r)$ means the sum of $2 + 3r$ from $r = 0$ to $r = 10$

$= 2 + 5 + 8 + \ldots + 32$

$\sum_{r=5}^{r=15} (10 - 2r)$ means the sum of $(10 - 2r)$ from $r = 5$ to $r = 15$

$= 0 + -2 + -4 + \ldots + -20$

Example 18

Calculate $\sum_{r=1}^{r=20} 4r + 1$

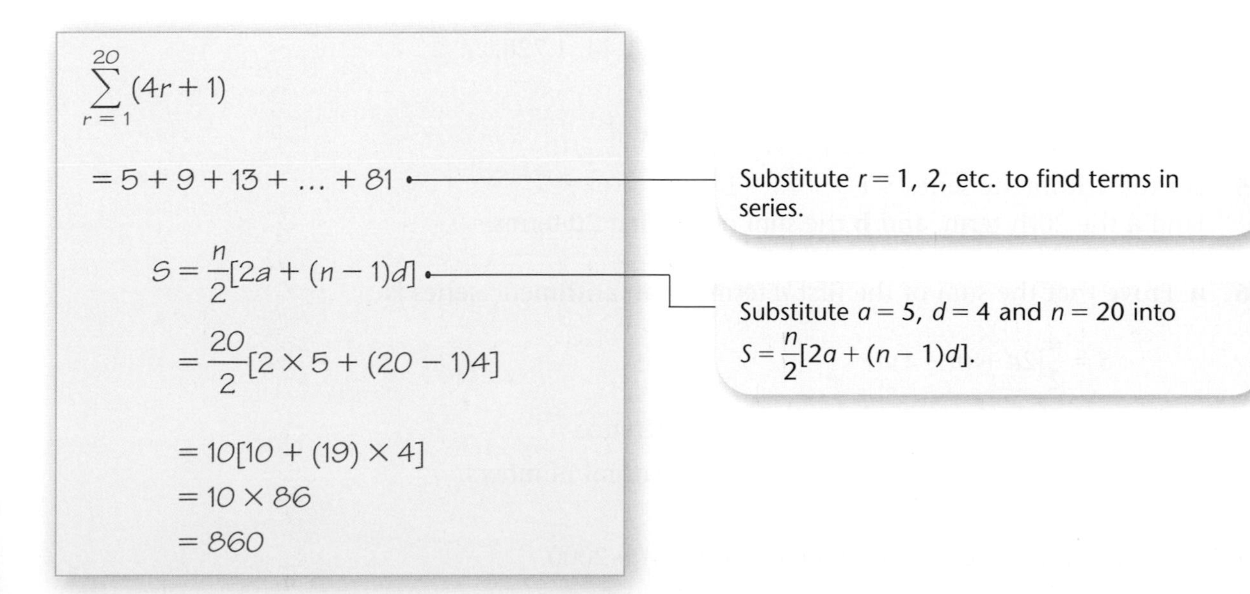

$\sum_{r=1}^{20} (4r + 1)$

$= 5 + 9 + 13 + \ldots + 81$

Substitute $r = 1, 2$, etc. to find terms in series.

$S = \frac{n}{2}[2a + (n - 1)d]$

Substitute $a = 5$, $d = 4$ and $n = 20$ into $S = \frac{n}{2}[2a + (n - 1)d]$.

$= \frac{20}{2}[2 \times 5 + (20 - 1)4]$

$= 10[10 + (19) \times 4]$

$= 10 \times 86$

$= 860$

Exercise 6G

1 Rewrite the following sums using Σ notation:

a $4 + 7 + 10 + \ldots + 31$

b $2 + 5 + 8 + 11 + \ldots + 89$

c $40 + 36 + 32 + \ldots + 0$

d The multiples of 6 less than 100

2 Calculate the following:

a $\sum_{r=1}^{5} 3r$

b $\sum_{r=1}^{10} (4r - 1)$

c $\sum_{r=1}^{20} (5r - 2)$

d $\sum_{r=0}^{5} r(r + 1)$

3 For what value of n does $\sum_{r=1}^{n} (5r + 3)$ first exceed 1000?

4 For what value of n would $\sum_{r=1}^{n} (100 - 4r) = 0$?

Mixed exercise 6H

1 The rth term in a sequence is $2 + 3r$. Find the first three terms of the sequence.

2 The rth term in a sequence is $(r + 3)(r - 4)$. Find the value of r for the term that has the value 78.

3 A sequence is formed from an inductive relationship:

$$U_{n+1} = 2U_n + 5$$

Given that $U_1 = 2$, find the first four terms of the sequence.

4 Find a rule that describes the following sequences:

a 5, 11, 17, 23, …

b 3, 6, 9, 12, …

c 1, 3, 9, 27, …

d 10, 5, 0, −5, …

e 1, 4, 9, 16, …

f 1, 1.2, 1.44, 1.728, …

Which of the above are arithmetic sequences?
For the ones that are, state the values of a and d.

5 For the arithmetic series $5 + 9 + 13 + 17 + \ldots$
Find **a** the 20th term, and **b** the sum of the first 20 terms.

6 **a** Prove that the sum of the first n terms in an arithmetic series is

$$S = \frac{n}{2}[2a + (n - 1)d]$$

where a = first term and d = common difference.

b Use this to find the sum of the first 100 natural numbers.

7 Find the least value of n for which $\sum_{r=1}^{n} (4r - 3) > 2000$.

8 A salesman is paid commission of £10 per week for each life insurance policy that he has sold. Each week he sells one new policy so that he is paid £10 commission in the first week, £20 commission in the second week, £30 commission in the third week and so on.

a Find his total commission in the first year of 52 weeks.

b In the second year the commission increases to £11 per week on new policies sold, although it remains at £10 per week for policies sold in the first year. He continues to sell one policy per week. Show that he is paid £542 in the second week of his second year.

c Find the total commission paid to him in the second year. *E*

9 The sum of the first two terms of an arithmetic series is 47.
The thirtieth term of this series is -62. Find:

a the first term of the series and the common difference

b the sum of the first 60 terms of the series. *E*

10 **a** Find the sum of the integers which are divisible by 3 and lie between 1 and 400.

b Hence, or otherwise, find the sum of the integers, from 1 to 400 inclusive, which are **not** divisible by 3. *E*

11 A polygon has 10 sides. The lengths of the sides, starting with the smallest, form an arithmetic series. The perimeter of the polygon is 675 cm and the length of the longest side is twice that of the shortest side. Find, for this series:

a the common difference

b the first term. *E*

12 A sequence of terms $\{U_n\}$ is defined for $n \geqslant 1$, by the recurrence relation $U_{n+2} = 2kU_{n+1} + 15U_n$, where k is a constant. Given that $U_1 = 1$ and $U_2 = -2$:

a find an expression, in terms of k, for U_3

b hence find an expression, in terms of k, for U_4

c given also that $U_4 = -38$, find the possible values of k. *E*

13 Prospectors are drilling for oil. The cost of drilling to a depth of 50 m is £500. To drill a further 50 m costs £640 and, hence, the total cost of drilling to a depth of 100 m is £1140. Each subsequent extra depth of 50 m costs £140 more to drill than the previous 50 m.

a Show that the cost of drilling to a depth of 500 m is £11 300.

b The total sum of money available for drilling is £76 000. Find, to the nearest 50 m, the greatest depth that can be drilled. *E*

14 Prove that the sum of the first $2n$ multiples of 4 is $4n(2n + 1)$. *E*

15 A sequence of numbers $\{U_n\}$ is defined, for $n \geqslant 1$, by the recurrence relation $U_{n+1} = kU_n - 4$, where k is a constant. Given that $U_1 = 2$:

a find expressions, in terms of k, for U_2 and U_3

b given also that $U_3 = 26$, use algebra to find the possible values of k. *E*

16 Each year, for 40 years, Anne will pay money into a savings scheme. In the first year she pays in £500. Her payments then increase by £50 each year, so that she pays in £550 in the second year, £600 in the third year, and so on.

a Find the amount that Anne will pay in the 40th year.

b Find the total amount that Anne will pay in over the 40 years.

c Over the same 40 years, Brian will also pay money into the savings scheme. In the first year he pays in £890 and his payments then increase by £d each year. Given that Brian and Anne will pay in exactly the same amount over the 40 years, find the value of d. **E**

17 The fifth term of an arithmetic series is 14 and the sum of the first three terms of the series is -3.

a Use algebra to show that the first term of the series is -6 and calculate the common difference of the series.

b Given that the nth term of the series is greater than 282, find the least possible value of n. **E**

18 The fourth term of an arithmetic series is $3k$, where k is a constant, and the sum of the first six terms of the series is $7k + 9$.

a Show that the first term of the series is $9 - 8k$.

b Find an expression for the common difference of the series in terms of k.

Given that the seventh term of the series is 12, calculate:

c the value of k

d the sum of the first 20 terms of the series. **E**

Summary of key points

1 A series of numbers following a set rule is called a sequence.

3, 7, 11, 15, 19, ... is an example of a sequence.

2 Each number in a sequence is called a term.

3 The *n*th term of a sequence is sometimes called the general term.

4 A sequence can be expressed as a formula for the *n*th term. For example the formula $U_n = 4n + 1$ produces the sequence 5, 9, 13, 17, ... by replacing n with 1, 2, 3, 4, etc in $4n + 1$.

5 A sequence can be expressed by a recurrence relationship. For example the same sequence 5, 9, 13, 17,... can be formed from $U_{n+1} = U_n + 4$, $U_1 = 5$. (U_1 must be given.)

6 A recurrence relationship of the form

$$U_{k+1} = U_k + n,\ k \geqslant 1 \quad n \in \mathbb{Z}$$

is called an arithmetic sequence.

7 All arithmetic sequences can be put in the form

a	$+ (a + d)$	$+ (a + 2d)$	$+ (a + 3d)$	$+ (a + 4d)$	$+ (a + 5d)$
↑	↑	↑	↑	↑	↑
1st term	2nd term	3rd term	4th term	5th term	6th term

8 The *n*th term of an arithmetic series is $a + (n - 1)d$, where a is the first term and d is the common difference.

9 The formula for the sum of an arithmetic series is

$$S_n = \frac{n}{2}[2a + (n - 1)d]$$

or $S_n = \frac{n}{2}(a + L)$

where a is the first term, d is the common difference, n is the number of terms and L is the last term in the series.

10 You can use Σ to signify 'sum of'. You can use Σ to write series in a more concise way

e.g. $\sum_{r=1}^{10} (5 + 2r) = 7 + 9 + \ldots + 25$

After completing this chapter you should be able to

1. estimate the gradient of a curve
2. calculate the gradient function, $\frac{dy}{dx}$ for simple functions
3. calculate the gradient of a curve at any point
4. find the equation of the tangent and normal to a curve at a specified point
5. calculate the second differential $\frac{d^2y}{dx^2}$.

Differentiation

Sir Isaac Newton who, along with Gottfried Liebniz, devised the laws of calculus.

Did you know?

Differential calculus is an important part of A level Mathematics and is widely used in many branches of Science, Engineering and Business. Understanding it will help you to sketch a function by finding the maximum and minimum values.

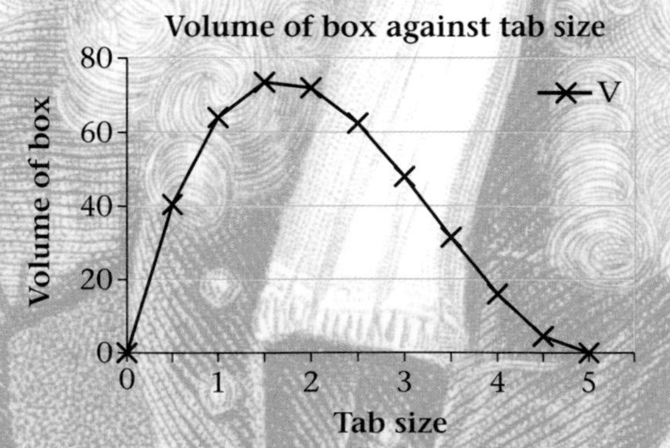

Differentiation enables us to find the exact value where the volume of the box described in Chapter 4 is maximised.

Successful businesses maximise profits and minimise costs.
A simple example to explain this might be a drinks manufacturer using cans that hold 330 ml. If the surface area of the can is as small as possible, then profits are maximised as the amount of aluminium used is minimised.

7.1 You can calculate an estimate of the gradient of a tangent.

In Section 5.1, you found the gradient of a **straight line** by calculation and by inspection of its equation.

The gradient of a curve changes as you move along it, and so:

■ **The gradient of a curve at a specific point is defined as being the same as the gradient of the tangent to the curve at that point.**

The tangent is a straight line, which touches, but does not cut, the curve. You cannot calculate the gradient of the tangent directly, as you know only one point on the tangent and you require two points to calculate the gradient of a line.

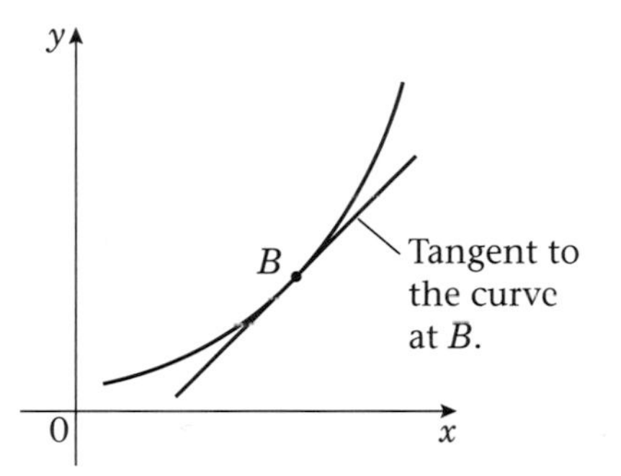

To find the gradient of the tangent at a point B on a curve with a known equation, you can find the gradient of chords joining B to other points close to B on the curve. You can then investigate the values of these gradients as the other points become closer to B. You should find the values become very close to a limiting value, which is the value of the gradient of the tangent, and is also the gradient of the curve at the point B.

Example 1

The points shown on the curve with equation $y = x^2$, are $O(0, 0)$, $A(\frac{1}{2}, \frac{1}{4})$, $B(1, 1)$, $C(1.5, 2.25)$ and $D(2, 4)$.

a Calculate the gradients of:

i OB

ii AB

iii BC

iv BD

b What do you deduce about the gradient of the tangent at the point B?

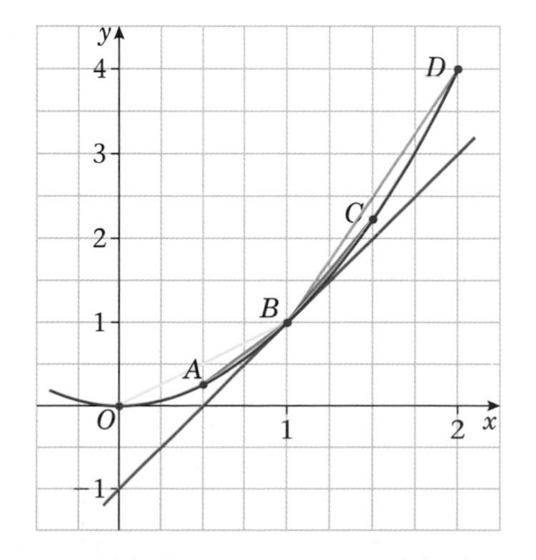

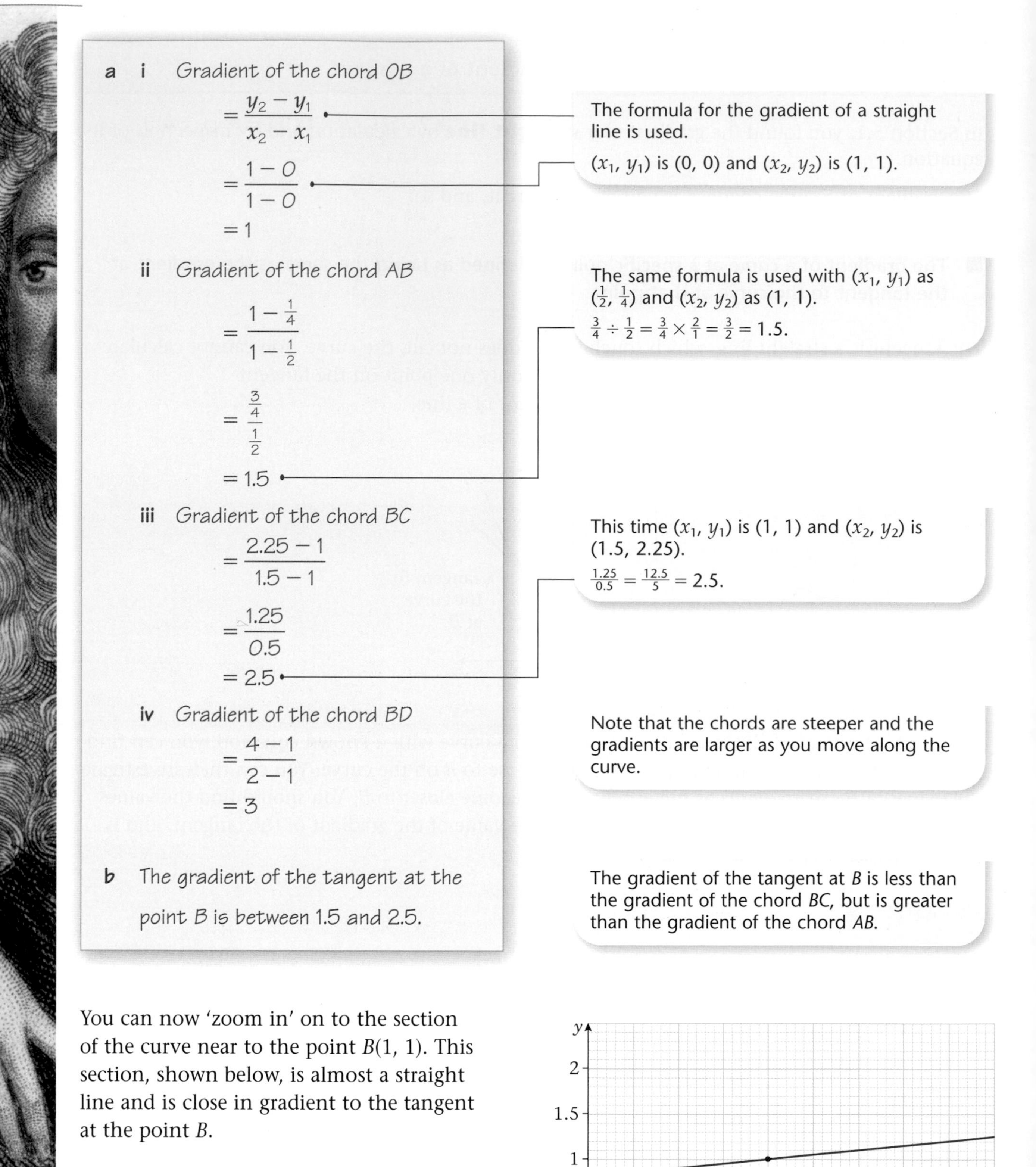

a i Gradient of the chord OB

$$= \frac{y_2 - y_1}{x_2 - x_1}$$

$$= \frac{1 - 0}{1 - 0}$$

$$= 1$$

The formula for the gradient of a straight line is used.

(x_1, y_1) is (0, 0) and (x_2, y_2) is (1, 1).

ii Gradient of the chord AB

$$= \frac{1 - \frac{1}{4}}{1 - \frac{1}{2}}$$

$$= \frac{\frac{3}{4}}{\frac{1}{2}}$$

$$= 1.5$$

The same formula is used with (x_1, y_1) as $(\frac{1}{2}, \frac{1}{4})$ and (x_2, y_2) as (1, 1).

$\frac{3}{4} \div \frac{1}{2} = \frac{3}{4} \times \frac{2}{1} = \frac{3}{2} = 1.5$.

iii Gradient of the chord BC

$$= \frac{2.25 - 1}{1.5 - 1}$$

$$= \frac{1.25}{0.5}$$

$$= 2.5$$

This time (x_1, y_1) is (1, 1) and (x_2, y_2) is (1.5, 2.25).

$\frac{1.25}{0.5} = \frac{12.5}{5} = 2.5$.

iv Gradient of the chord BD

$$= \frac{4 - 1}{2 - 1}$$

$$= 3$$

Note that the chords are steeper and the gradients are larger as you move along the curve.

b The gradient of the tangent at the point B is between 1.5 and 2.5.

The gradient of the tangent at B is less than the gradient of the chord BC, but is greater than the gradient of the chord AB.

You can now 'zoom in' on to the section of the curve near to the point $B(1, 1)$. This section, shown below, is almost a straight line and is close in gradient to the tangent at the point B.

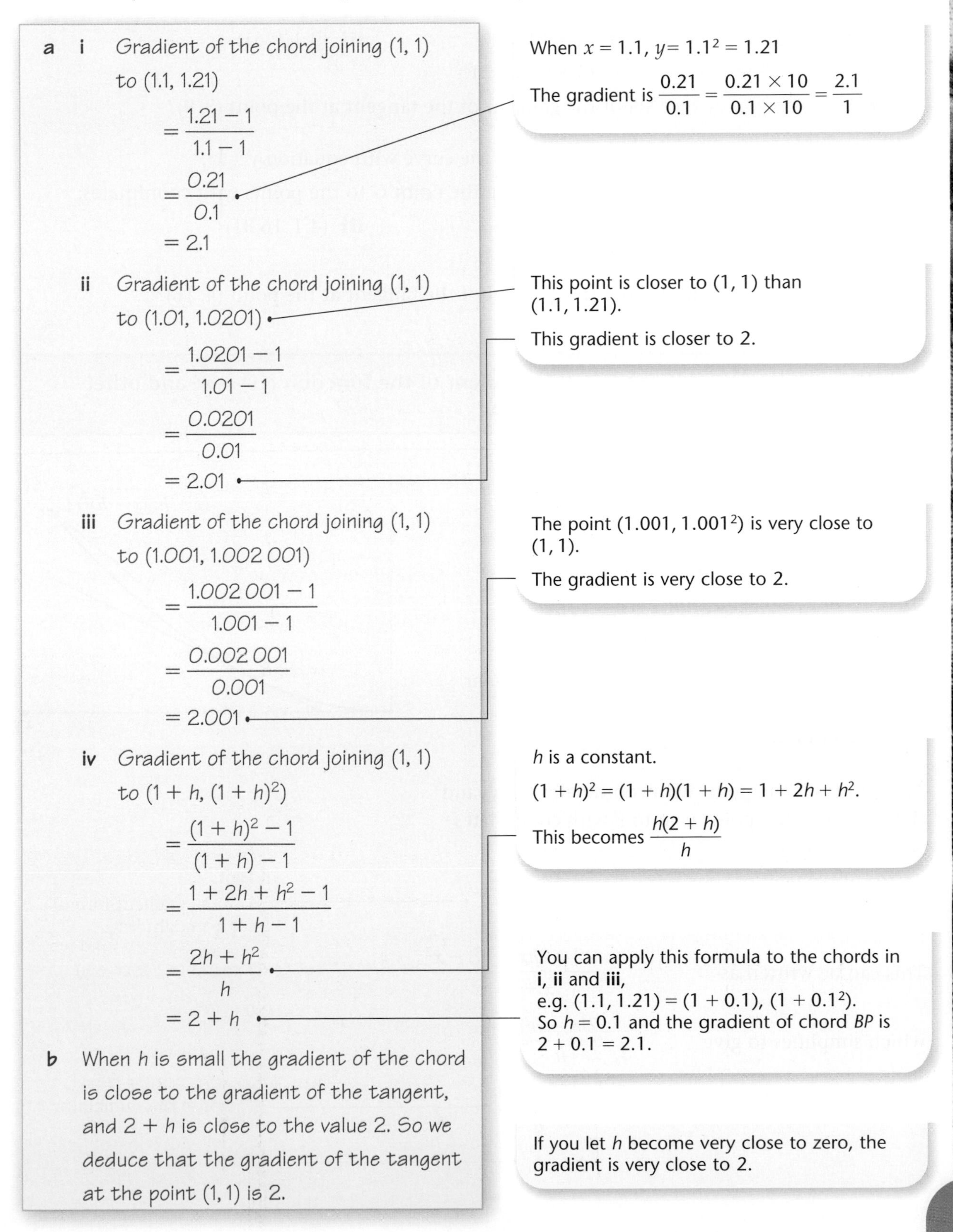

Example 2

a For the same curve as Example 1, find the gradient of the chord BP when P has coordinates:
i $(1.1, 1.21)$ **ii** $(1.01, 1.0201)$ **iii** $(1.001, 1.002\,001)$ **iv** $(1 + h, (1 + h)^2)$

b What do you deduce about the gradient of the tangent at the point B?

a i Gradient of the chord joining (1, 1) to (1.1, 1.21)

$$= \frac{1.21 - 1}{1.1 - 1}$$

$$= \frac{0.21}{0.1}$$

$$= 2.1$$

When $x = 1.1$, $y = 1.1^2 = 1.21$

The gradient is $\frac{0.21}{0.1} = \frac{0.21 \times 10}{0.1 \times 10} = \frac{2.1}{1}$

ii Gradient of the chord joining (1, 1) to (1.01, 1.0201)

$$= \frac{1.0201 - 1}{1.01 - 1}$$

$$= \frac{0.0201}{0.01}$$

$$= 2.01$$

This point is closer to (1, 1) than (1.1, 1.21).

This gradient is closer to 2.

iii Gradient of the chord joining (1, 1) to (1.001, 1.002 001)

$$= \frac{1.002\,001 - 1}{1.001 - 1}$$

$$= \frac{0.002\,001}{0.001}$$

$$= 2.001$$

The point $(1.001, 1.001^2)$ is very close to (1, 1).

The gradient is very close to 2.

iv Gradient of the chord joining (1, 1) to $(1 + h, (1 + h)^2)$

$$= \frac{(1 + h)^2 - 1}{(1 + h) - 1}$$

$$= \frac{1 + 2h + h^2 - 1}{1 + h - 1}$$

$$= \frac{2h + h^2}{h}$$

$$= 2 + h$$

h is a constant.

$(1 + h)^2 = (1 + h)(1 + h) = 1 + 2h + h^2$.

This becomes $\frac{h(2 + h)}{h}$

You can apply this formula to the chords in **i**, **ii** and **iii**,
e.g. $(1.1, 1.21) = (1 + 0.1), (1 + 0.1^2)$.
So $h = 0.1$ and the gradient of chord BP is $2 + 0.1 = 2.1$.

b When h is small the gradient of the chord is close to the gradient of the tangent, and $2 + h$ is close to the value 2. So we deduce that the gradient of the tangent at the point (1, 1) is 2.

If you let h become very close to zero, the gradient is very close to 2.

Exercise 7A

Questions like these will not appear in the examination papers.

1 F is the point with co-ordinates (3, 9) on the curve with equation $y = x^2$.

a Find the gradients of the chords joining the point F to the points with coordinates:

i (4, 16) **ii** (3.5, 12.25) **iii** (3.1, 9.61)

iv (3.01, 9.0601) **v** $(3 + h, (3 + h)^2)$

b What do you deduce about the gradient of the tangent at the point (3, 9)?

2 G is the point with coordinates (4, 16) on the curve with equation $y = x^2$.

a Find the gradients of the chords joining the point G to the points with coordinates:

i (5, 25) **ii** (4.5, 20.25) **iii** (4.1, 16.81)

iv (4.01, 16.0801) **v** $(4 + h, (4 + h)^2)$

b What do you deduce about the gradient of the tangent at the point (4, 16)?

7.2 You can find the formula for the gradient of the function $f(x) = x^2$ and other functions of the form $f(x) = x^n$, $n \in \mathbb{R}$.

Examples 2 to 4 show you how to derive the formulae and will not be tested.

In the following sketch, the gradient of the tangent $y = f(x)$ at a point B is found by starting with the gradient of a chord BC.

- **The gradient of the tangent at any particular point is the rate of change of y with respect to x.**

The point B is the point with coordinates (x, x^2) and the point C is the point near to B with coordinates $(x + h, (x + h^2))$.

The gradient of the chord BC is $\dfrac{(x + h)^2 - x^2}{(x + h) - x}$

This can be written as $\dfrac{(x^2 + 2hx + h^2) - x^2}{x + h - x}$

which simplifies to give $\dfrac{2hx + h^2}{h}$

$= \dfrac{h(2x + h)}{h}$

$= 2x + h.$

Hint:
Use the gradient formula for a straight line.
Expand $(x + h)(x + h)$.
Factorise the numerator.
Cancel the factor h.

As h becomes smaller the gradient of the chord becomes closer to the gradient of the tangent to the curve at the point B.

The gradient of the tangent at the point B to the curve with equation $y = x^2$ is therefore given by the formula: gradient $= 2x$.

In general you will find that the gradients of the tangents to a given curve can be expressed by a formula related to the equation of the curve.

■ **The gradient formula for $y = \mathrm{f}(x)$ is given by the equation: gradient $= \mathrm{f}'(x)$, where $\mathrm{f}'(x)$ is called the derived function.**

$\mathrm{f}'(x)$ is defined as the gradient of the curve $y = \mathrm{f}(x)$ at the general point $(x, \mathrm{f}(x))$. It is also the gradient of the tangent to the curve at that point.

So far you have seen that when $\mathrm{f}(x) = x^2$, $\mathrm{f}'(x) = 2x$.

You can use this result to determine the gradient of the curve $y = x^2$ at any specified point on the curve.

You can also use a similar approach to establish a gradient formula for the graph of $y = \mathrm{f}(x)$, where $\mathrm{f}(x)$ is a power of x, i.e. $\mathrm{f}(x) = x^n$, where n is any real number.

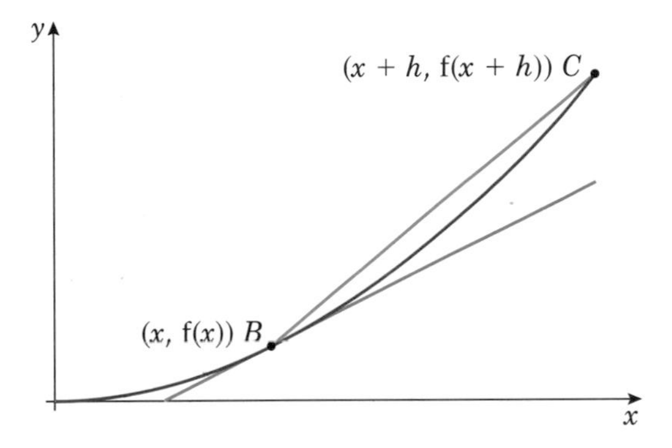

Again you need to consider the gradient of a chord joining two points which are close together on the curve and determine what happens when the points become very close together.

This time the point B has coordinates $(x, \mathrm{f}(x))$ and the point C is the point near to B with coordinates $(x + h, \mathrm{f}(x + h))$.

The gradient of BC is

$$\frac{\mathrm{f}(x + h) - \mathrm{f}(x)}{(x + h) - x}$$

So as h becomes small and the gradient of the chord becomes close to the gradient of the tangent, the definition of $\mathrm{f}'(x)$ is given as

$$\lim_{h \to 0}\left[\frac{\mathrm{f}(x + h) - \mathrm{f}(x)}{h}\right]$$

Using this definition you can differentiate a function of the form $\mathrm{f}(x) = x^n$.

Example 3

Find, from the definition of the derived function, an expression for $f'(x)$ when $f(x) = x^3$.

$$f'(x) = \lim_{h \to 0} \frac{f(x+h) - f(x)}{(x+h) - x}$$

$$= \lim_{h \to 0} \frac{(x+h)^3 - (x)^3}{(x+h) - x}$$

$$= \lim_{h \to 0} \frac{x^3 + 3x^2h + 3xh^2 + h^3 - x^3}{(x+h) - x}$$

$(x + h)^3 = (x + h)(x + h)^2$
$= (x + h)(x^2 + 2hx + h^2)$
which expands to give
$x^3 + 3x^2h + 3xh^2 + h^3$.

$$= \lim_{h \to 0} \frac{3x^2h + 3xh^2 + h^3}{h}$$

$$= \lim_{h \to 0} \frac{h(3x^2 + 3xh + h^2)}{h}$$

Factorise the numerator.

$$= \lim_{h \to 0} 3x^2 + 3xh + h^2$$

The $3xh$ term and the h^2 term become zero.

As $h \to 0$ the limiting value is $3x^2$.

So when $f(x) = x^3$, $f'(x) = 3x^2$.

Example 4

Find, from the definition of the derived function, an expression for $f'(x)$ when $f(x) = \frac{1}{x}$.

$$f'(x) = \lim_{h \to 0} \frac{f(x+h) - f(x)}{(x+h) - x}$$

$$= \lim_{h \to 0} \frac{\frac{1}{(x+h)} - \frac{1}{x}}{(x+h) - x}$$

$$= \lim_{h \to 0} \frac{\frac{x - (x+h)}{x(x+h)}}{(x+h) - x}$$

Use a common denominator.

$$= \lim_{h \to 0} \frac{-h}{x(x+h)} \div h$$

A fraction over a denominator h is the same as the fraction divided by h, and the h then cancels.

$$= \lim_{h \to 0} -\frac{1}{x^2 + xh}$$

The xh term becomes zero.

As $h \to 0$ the limiting value is $-\frac{1}{x^2} = -x^{-2}$.

So when $f(x) = x^{-1}$, $f'(x) = (-1)x^{-2}$.

You have found that:

$f(x) = x^2$ gives $f'(x) = 2x^{2-1}$

$f(x) = x^3$ gives $f'(x) = 3x^{3-1}$

$f(x) = x^{-1}$ gives $f'(x) = -1x^{-1-1}$

Hint: Notice the pattern in these results is the same each time.

Also, you know that the gradient of the straight line $y = x$ is 1, and that the gradient of the straight line $y = 1$ is 0.

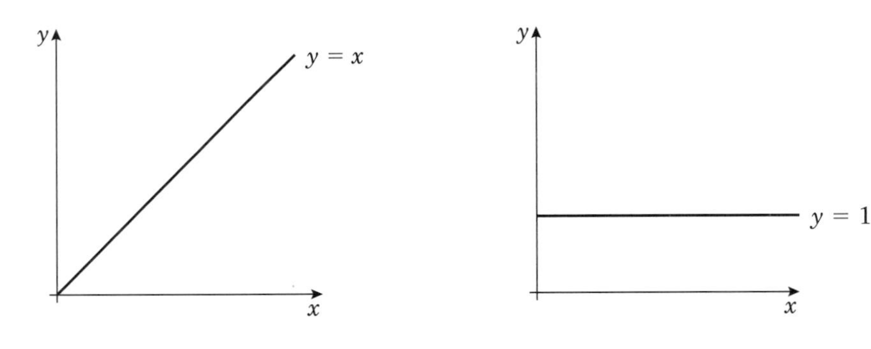

So $f(x) = x^1$ gives $f'(x) = 1x^{1-1}$

and $f(x) = x^0$ gives $f'(x) = 0x^{0-1}$

Hint: Notice the same pattern for these linear functions.

■ **In general it can be shown that if**

$$f(x) = x^n,\ n \in \mathbb{R} \text{ then } f'(x) = nx^{n-1}$$

So the original power multiplies the expression and the power of x is reduced by 1.

Example 5

Find the derived function when $f(x)$ equals:

a x^6 **b** $x^{\frac{1}{2}}$ **c** x^{-2} **d** $\frac{x}{x^5}$ **e** $x^2 \times x^3$

a $6x^5$

The power 6 is reduced to power 5 and the 6 multiplies the answer.

b $f(x) = x^{\frac{1}{2}}$

$f'(x) = \frac{1}{2}x^{-\frac{1}{2}}$

$= \frac{1}{2\sqrt{x}}$

The power $\frac{1}{2}$ is reduced to $\frac{1}{2} - 1 = -\frac{1}{2}$, and the $\frac{1}{2}$ multiplies the answer. This is then rewritten in an alternative form.

c $f(x) = x^{-2}$

$f'(x) = -2x^{-3}$

$= -\frac{2}{x^3}$

The power -2 is reduced to -3 and the -2 multiplies the answer. This is also rewritten in an alternative form using knowledge of negative powers.

d Let $f(x) = x \div x^5$
$= x^{-4}$

Simplify using rules of powers to give one simple power, i.e. subtract $1 - 5 = -4$.

So $f'(x) = -4x^{-5}$
$= -\dfrac{4}{x^5}$

Reduce the power -4 to give -5, then multiply your answer by -4.

e Let $f(x) = x^2 \times x^3$
$= x^5$

Add the powers this time to give $2 + 3 = 5$.

So $f'(x) = 5x^4$

Reduce the power 5 to 4 and multiply your answer by 5.

Exercise 7B

Find the derived function, given that f(x) equals:

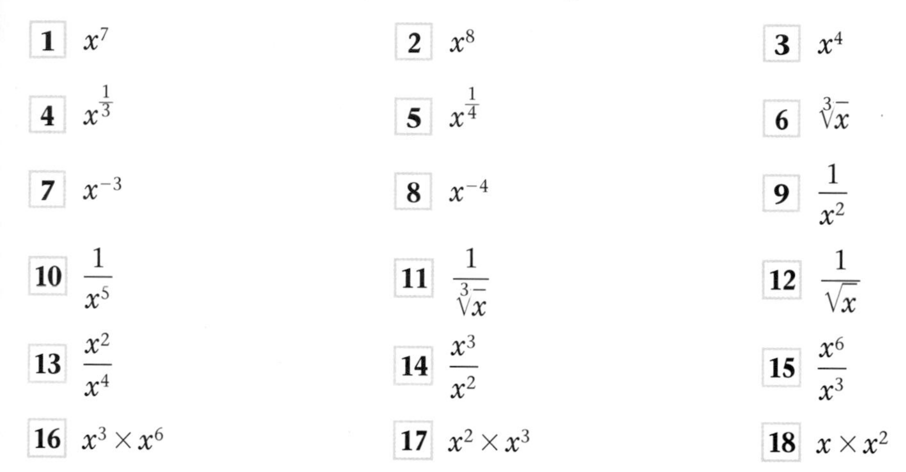

1 x^7 **2** x^8 **3** x^4

4 $x^{\frac{1}{3}}$ **5** $x^{\frac{1}{4}}$ **6** $\sqrt[3]{x}$

7 x^{-3} **8** x^{-4} **9** $\dfrac{1}{x^2}$

10 $\dfrac{1}{x^5}$ **11** $\dfrac{1}{\sqrt[3]{x}}$ **12** $\dfrac{1}{\sqrt{x}}$

13 $\dfrac{x^2}{x^4}$ **14** $\dfrac{x^3}{x^2}$ **15** $\dfrac{x^6}{x^3}$

16 $x^3 \times x^6$ **17** $x^2 \times x^3$ **18** $x \times x^2$

7.3 You can find the gradient formula for a function such as $f(x) = 4x^2 - 8x + 3$ and other functions of the form $f(x) = ax^2 + bx + c$, where a, b and c are constants.

You can use an alternative notation when finding the gradient function.

Again, you find the gradient of the tangent at a point B by starting with the gradient of a chord BC. This time the point B is the point with coordinates (x, y) and the point C is the point near to B with coordinates $(x + \delta x, y + \delta y)$. δx is called delta x and is a single symbol which stands for a small change in the value of x. This was denoted by h in Section 7.2. Also δy is called 'delta y' and is a single symbol which stands for a small change in the value of y.

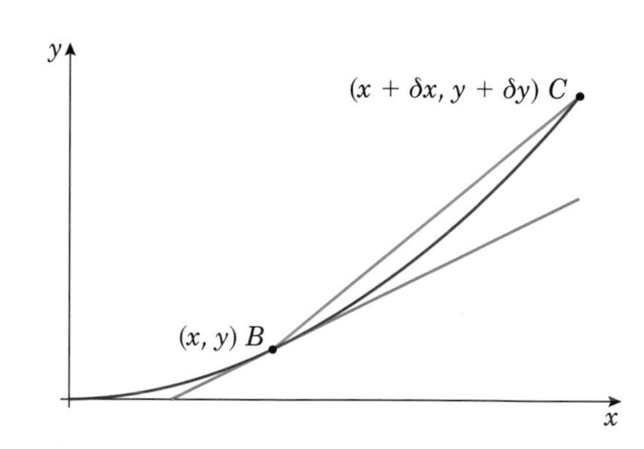

The gradient of the chord BC is then

$$\frac{y + \delta y - y}{x + \delta x - x} = \frac{\delta y}{\delta x}$$

But both B and C lie on the curve with equation $y = \mathrm{f}(x)$ and so B is the point $(x, \mathrm{f}(x))$ and C is the point $(x + \delta x, \mathrm{f}(x + \delta x))$.

So the gradient of BC can also be written as

$$\frac{\mathrm{f}(x + \delta x) - \mathrm{f}(x)}{(x + \delta x) - x} = \frac{\mathrm{f}(x + \delta x) - \mathrm{f}(x)}{\delta x}$$

You can make the value of δx very small and you will find that the smaller the value of δx, the smaller the value of δy will be.
The limiting value of the gradient of the chord is the gradient of the tangent at B, which is also the gradient of the curve at B.

This is called the rate of change of y with respect to x at the point B and is denoted by $\frac{\mathrm{d}y}{\mathrm{d}x}$.

$$\frac{\mathrm{d}y}{\mathrm{d}x} = \lim_{\delta x \to 0}\left(\frac{\delta y}{\delta x}\right)$$

$$= \lim_{\delta x \to 0} \frac{\mathrm{f}(x + \delta x) - \mathrm{f}(x)}{\delta x}$$

$\frac{\mathrm{d}y}{\mathrm{d}x}$ is called the derivative of y with respect to x.

Also $\frac{\mathrm{d}y}{\mathrm{d}x} = \mathrm{f}'(x)$.

The process of finding $\frac{\mathrm{d}y}{\mathrm{d}x}$ when y is given is called differentiation.

■ **When $y = x^n$, $\frac{\mathrm{d}y}{\mathrm{d}x} = nx^{n-1}$ for all real values of n.**

You can also differentiate the general quadratic equation $y = ax^2 + bx + c$.

Using the definition that $\frac{\mathrm{d}y}{\mathrm{d}x} = \lim_{\delta x \to 0} \frac{\mathrm{f}(x + \delta x) - \mathrm{f}(x)}{\delta x}$

Then $\frac{\mathrm{d}y}{\mathrm{d}x} = \lim_{\delta x \to 0} \frac{a(x + \delta x)^2 + b(x + \delta x) + c - (ax^2 + bx + c)}{x + \delta x - x}$

$$= \lim_{\delta x \to 0} \frac{2ax\delta x + a(\delta x)^2 + b\delta x}{\delta x}$$

$$= 2ax + b$$

Therefore when $y = ax^2 + bx + c$, $\frac{\mathrm{d}y}{\mathrm{d}x} = 2ax + b$.

Hint:
Factorise the numerator to give.
$\delta x(2ax + a\delta x + b)$
then simplify the fraction as δx is a common factor.

$a\delta x$ term becomes zero.

Consider the three sketches below:

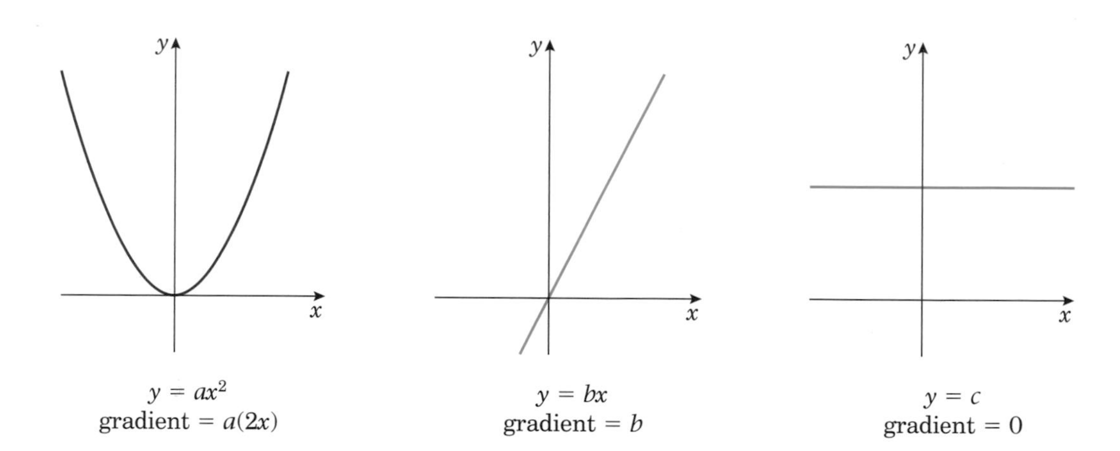

Combining these functions gives $y = ax^2 + bx + c$, with gradient given by $\frac{dy}{dx} = 2ax + b$.

Example 6

Find $\frac{dy}{dx}$ when y equals:

a x^2 **b** 4 **c** $12x + 3$ **d** $x^2 - 6x - 4$ **e** $3 - 5x^2$

	Working	Notes
a	$2x$	
b	0	The line $y = 4$ has zero gradient.
c	12	Using $y = mx + c$, the gradient is the value of m.
d	$2x - 6$	Use the result given above with $a = 1$, $b = -6$, $c = -4$.
e	$-10x$	$a = -5$, $b = 0$ and $c = 3$.

Example 7

Let $f(x) = 4x^2 - 8x + 3$.

a Find the gradient of $y = f(x)$ at the point $(\frac{1}{2}, 0)$.

b Find the coordinates of the point on the graph of $y = f(x)$ where the gradient is 8.

c Find the gradient of $y = f(x)$ at the points where the curve meets the line $y = 4x - 5$.

Working	Notes
a As $y = 4x^2 - 8x + 3$	
$\frac{dy}{dx} = f'(x) = 8x - 8 + 0$	First find $f'(x)$, the derived function, then substitute the x-coordinate value to obtain the gradient.
So $f'(\frac{1}{2}) = -4$	

b $\frac{dy}{dx} = f'(x) = 8x - 8 + 0 = 8$

So $x = 2$

So $y = f(2) = 3$

The point where the gradient is 8 is $(2, 3)$.

Put the gradient function equal to 8. Then solve the equation you have obtained to give the value of x.

Substitute this value for x into $f(x)$ to give the value of y and interpret your answer in words.

c $4x^2 - 8x + 3 = 4x - 5$

$4x^2 - 12x + 8 = 0$

$x^2 - 3x + 2 = 0$

$(x - 2)(x - 1) = 0$

So $x = 1$ or $x = 2$

At $x = 1$ the gradient is 0.

At $x = 2$ the gradient is 8, as in part **b**.

Put $f(x) = 4x - 5$, then rearrange and collect terms to give a quadratic equation.

Divide by the common factor 4.

Solve the quadratic equation by factorising, or by using the quadratic formula

$$x = \frac{-b \pm \sqrt{(b^2 - 4ac)}}{2a}$$

Substitute the values of x into $f'(x) = 8x - 8$ to give the gradients at the specified points.

Exercise 7C

1 Find $\frac{dy}{dx}$ when y equals:

a $2x^2 - 6x + 3$

b $\frac{1}{2}x^2 + 12x$

c $4x^2 - 6$

d $8x^2 + 7x + 12$

e $5 + 4x - 5x^2$

2 Find the gradient of the curve whose equation is

a $y = 3x^2$ at the point $(2, 12)$

b $y = x^2 + 4x$ at the point $(1, 5)$

c $y = 2x^2 - x - 1$ at the point $(2, 5)$

d $y = \frac{1}{2}x^2 + \frac{3}{2}x$ at the point $(1, 2)$

e $y = 3 - x^2$ at the point $(1, 2)$

f $y = 4 - 2x^2$ at the point $(-1, 2)$

3 Find the y-coordinate and the value of the gradient at the point P with x-coordinate 1 on the curve with equation $y = 3 + 2x - x^2$.

4 Find the coordinates of the point on the curve with equation $y = x^2 + 5x - 4$ where the gradient is 3.

5 Find the gradients of the curve $y = x^2 - 5x + 10$ at the points A and B where the curve meets the line $y = 4$.

6 Find the gradients of the curve $y = 2x^2$ at the points C and D where the curve meets the line $y = x + 3$.

7.4 You can find the gradient formula for a function such as $f(x) = x^3 + x^2 - x^{\frac{1}{2}}$ where the powers of x are real numbers $a_n x^n + a_{n-1}x^{n-1} + \ldots + a_0$, where $a_n, a_{n-1}, \ldots, a_0$ are constants, $a_n \neq 0$ and $n \in \mathbb{R}$.

You know that if $y = x^n$, then $\frac{dy}{dx} = nx^{n-1}$.

This is true for all real values of n.

It can also be shown that

■ **if $y = ax^n$. where a is a constant then $\frac{dy}{dx} = anx^{n-1}$.**

Hint: Note that you again reduce the power by 1 and the original power multiplies the expression.

Also

■ **if $y = f(x) \pm g(x)$ then $\frac{dy}{dx} = f'(x) \pm g'(x)$.**

These standard results can be assumed without proof at A Level.

Example 8

Use standard results to differentiate:

a $x^3 + x^2 - x^{\frac{1}{2}}$ **b** $2x^{-3}$ **c** $\frac{1}{3}x^{\frac{1}{2}} + 4x^2$

a $y = x^3 + x^2 - x^{\frac{1}{2}}$

So $\frac{dy}{dx} = 3x^2 + 2x - \frac{1}{2}x^{-\frac{1}{2}}$

Differentiate each term as you come to it. First x^3, then x^2, then $-x^{\frac{1}{2}}$.

b $y = 2x^{-3}$

So $\frac{dy}{dx} = -6x^{-4}$

$= -\frac{6}{x^4}$

Differentiate x^{-3}, then multiply the answer by 2.

c $x = \frac{1}{3}x^{\frac{1}{2}} + 4x^2$

So $\frac{dy}{dx} = \frac{1}{3} \times \frac{1}{2}x^{-\frac{1}{2}} + 8x$

$= \frac{1}{6} \times x^{-\frac{1}{2}} + 8x$

Take each term as you come to it, and treat each term as a multiple.

Exercise 7D

1 Use standard results to differentiate:

a $x^4 + x^{-1}$ **b** $\frac{1}{2}x^{-2}$ **c** $2x^{-\frac{1}{2}}$

2 Find the gradient of the curve with equation $y = f(x)$ at the point A where:

a $f(x) = x^3 - 3x + 2$ and A is at $(-1, 4)$

b $f(x) = 3x^2 + 2x^{-1}$ and A is at $(2, 13)$

3 Find the point or points on the curve with equation $y = f(x)$, where the gradient is zero:

a $f(x) = x^2 - 5x$

b $f(x) = x^3 - 9x^2 + 24x - 20$

c $f(x) = x^{\frac{3}{2}} - 6x + 1$

d $f(x) = x^{-1} + 4x$

7.5 You can expand or simplify polynomial functions so that they are easier to differentiate.

Example 9

Use standard results to differentiate:

a $\dfrac{1}{4\sqrt{x}}$

b $x^3(3x + 1)$

c $\dfrac{x - 2}{x^2}$

a Let $y = \dfrac{1}{4\sqrt{x}}$

$= \frac{1}{4}x^{-\frac{1}{2}}$

Express the 4 in the denominator as a multiplier of $\frac{1}{4}$ and express the x term as power $-\frac{1}{2}$.

Therefore $\dfrac{dy}{dx} = -\frac{1}{8}x^{-\frac{3}{2}}$

Then differentiate by reducing the power of x and multiplying $\frac{1}{4}$ by $-\frac{1}{2}$.

b Let $y = x^3(3x + 1)$

$= 3x^4 + x^3$

Multiply out the brackets to give a polynomial function.

Therefore $\dfrac{dy}{dx} = 12x^3 + 3x^2$

Differentiate each term.

$= 3x^2(4x + 1)$

c Let $y = \dfrac{x - 2}{x^2}$

$= \dfrac{1}{x} - \dfrac{2}{x^2}$

Express the single fraction as two separate fractions, and simplify $\dfrac{x}{x^2}$ as $\dfrac{1}{x}$.

$= x^{-1} - 2x^{-2}$

Therefore $\dfrac{dy}{dx} = -x^{-2} + 4x^{-3}$

Then express the rational expressions as negative powers of x, and differentiate.

$= -\dfrac{1}{x^2} + \dfrac{4}{x^3}$

$= \dfrac{-(x - 4)}{x^3}$

Simplify by using a common denominator.

Exercise 7E

1 Use standard results to differentiate:

a $2\sqrt{x}$ **b** $\frac{3}{x^2}$ **c** $\frac{1}{3x^3}$

d $\frac{1}{3}x^3(x-2)$ **e** $\frac{2}{x^3}+\sqrt{x}$ **f** $\sqrt[3]{x}+\frac{1}{2x}$

g $\frac{2x+3}{x}$ **h** $\frac{3x^2-6}{x}$ **i** $\frac{2x^3+3x}{\sqrt{x}}$

j $x(x^2-x+2)$ **k** $3x^2(x^2+2x)$ **l** $(3x-2)\left(4x+\frac{1}{x}\right)$

2 Find the gradient of the curve with equation $y = \mathrm{f}(x)$ at the point A where:

a $\mathrm{f}(x) = x(x+1)$ and A is at $(0, 0)$ **b** $\mathrm{f}(x) = \frac{2x-6}{x^2}$ and A is at $(3, 0)$

c $\mathrm{f}(x) = \frac{1}{\sqrt{x}}$ and A is at $(\frac{1}{4}, 2)$ **d** $\mathrm{f}(x) = 3x - \frac{4}{x^2}$ and A is at $(2, 5)$

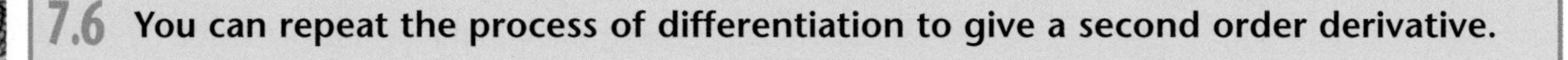

7.6 You can repeat the process of differentiation to give a second order derivative.

■ **A second order derivative is written as $\frac{d^2y}{dx^2}$, or $\mathrm{f}''(x)$ using function notation.**

Example 10

Given that $y = 3x^5 + \frac{4}{x^2}$ find:

a $\frac{dy}{dx}$ **b** $\frac{d^2y}{dx^2}$

a $y = 3x^5 + \frac{4}{x^2}$

$= 3x^5 + 4x^{-2}$ — Express the fraction as a negative power.

So $\frac{dy}{dx} = 15x^4 - 8x^{-3}$ — Differentiate a first time.

$= 15x^4 - \frac{8}{x^3}$

b $\frac{d^2y}{dx^2} = 60x^3 + 24x^{-4}$ — Differentiate a second time.

$= 60x^3 + \frac{24}{x^4}$

Example 11

Given that $f(x) = 3\sqrt{x} + \dfrac{1}{2\sqrt{x}}$, find:

a $f'(x)$ **b** $f''(x)$

a $f(x) = 3\sqrt{x} + \dfrac{1}{2\sqrt{x}}$

$= 3x^{\frac{1}{2}} + \frac{1}{2}x^{-\frac{1}{2}}$

$f'(x) = \frac{3}{2}x^{-\frac{1}{2}} - \frac{1}{4}x^{-\frac{3}{2}}$

b $f''(x) = -\frac{3}{4}x^{-\frac{3}{2}} + \frac{3}{8}x^{-\frac{5}{2}}$

Express the roots as fractional powers.

Multiply 3 by a half and reduce power of x.

Multiply a half by negative a half and reduce power of x.

Note that $\frac{1}{4} \times \frac{3}{2} = \frac{3}{8}$ and the product of two negatives is positive.

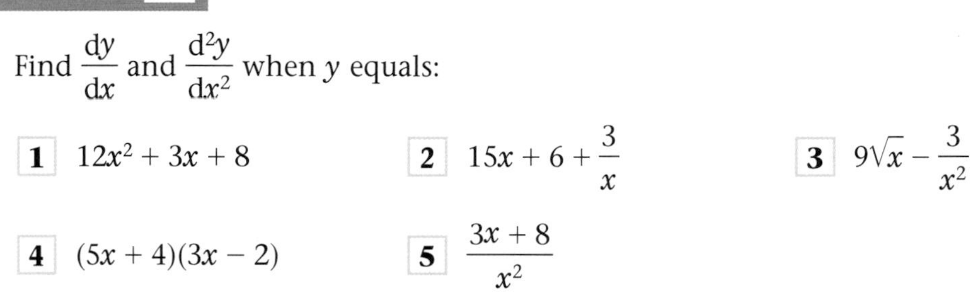

Exercise 7F

Find $\dfrac{dy}{dx}$ and $\dfrac{d^2y}{dx^2}$ when y equals:

1 $12x^2 + 3x + 8$

2 $15x + 6 + \dfrac{3}{x}$

3 $9\sqrt{x} - \dfrac{3}{x^2}$

4 $(5x + 4)(3x - 2)$

5 $\dfrac{3x + 8}{x^2}$

7.7 You can find the rate of change of a function f at a particular point by using $f'(x)$ and substituting in the value of x.

The variables in the relationship $y = f(x)$ are such that x is the independent variable and y is the dependent variable.
These variables often stand for quantities, where it is more meaningful to use letters, other than x and y, to suggest what these quantities are.

For example, it is usual to substitute t for time, V for volume, P for population, A for area, r for radius, s for displacement, h for height, v for velocity, θ for temperature, etc.

So $\dfrac{dV}{dt}$ might represent the gradient in a graph of volume against time. It therefore would represent the rate of change of volume with respect to time.

Also $\dfrac{dA}{dr}$ might represent the gradient in a graph of area against radius. It therefore would represent the rate of change of area with respect to radius.

You should know that the rate of change of velocity with respect to time is acceleration, and that the rate of change of displacement with respect to time is velocity.

Example 12

Given that the volume (V cm^3) of an expanding sphere is related to its radius (r cm) by the formula $V = \frac{4}{3}\pi r^3$, find the rate of change of volume with respect to radius at the instant when the radius is 5 cm.

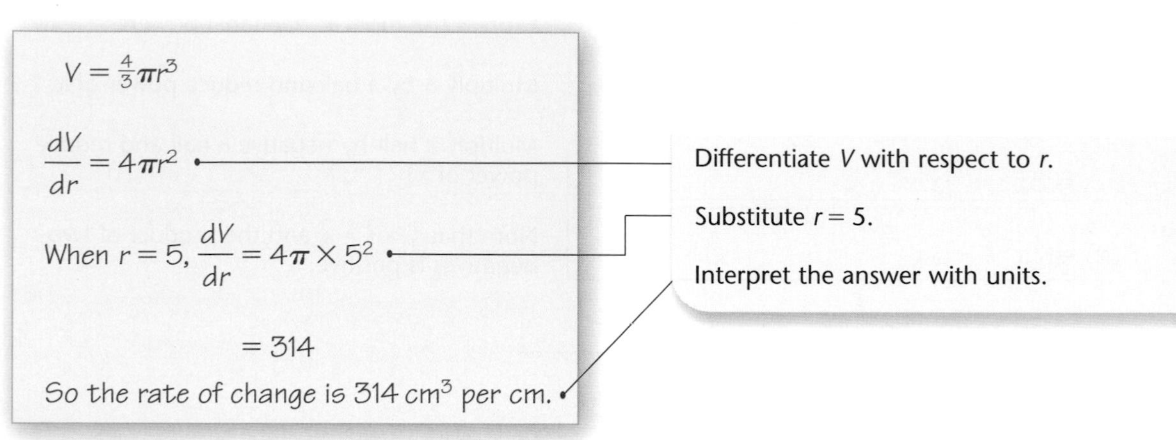

Exercise 7G

1 Find $\frac{d\theta}{dt}$ where $\theta = t^2 - 3t$

2 Find $\frac{dA}{dr}$ where $A = 2\pi r$

3 Given that $r = \frac{12}{t}$, find the value of $\frac{dr}{dt}$ when $t = 3$.

4 The surface area, A cm^2, of an expanding sphere of radiuis r cm is given by $A = 4\pi r^2$. Find the rate of change of the area with respect to the radius at the instant when the radius is 6 cm.

5 The displacement, s metres, of a car from a fixed point at time t seconds is given by $s = t^2 + 8t$. Find the rate of change of the displacement with respect to time at the instant when $t = 5$.

7.8 You can use differentiation to find the gradient of a tangent to a curve and you can then find the equation of the tangent and normal to that curve at a specified point.

The tangent at the point A $(a, f(a))$ has gradient $f'(a)$. You can use the formula for the equation of a straight line, $y - y_1 = m(x - x_1)$, to obtain the equation of the tangent at $(a, f(a))$.

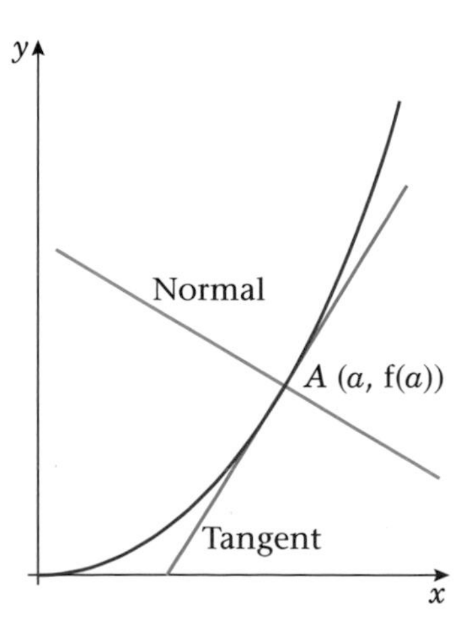

■ **The equation of the tangent to a curve at a point $(a, f(a))$ is $y - f(a) = f'(a)(x - a)$.**

The normal to the curve at the point A is defined as being the straight line through A which is perpendicular to the tangent at A (see sketch alongside).

The gradient of the normal is $-\dfrac{1}{f'(a)}$, because the product of the gradients of lines which are at right angles is -1.

■ **The equation of the normal at point A is $y - f(a) = -\dfrac{1}{f'(a)}(x - a)$.**

Example 13

Find the equation of the tangent to the curve $y = x^3 - 3x^2 + 2x - 1$ at the point (3, 5).

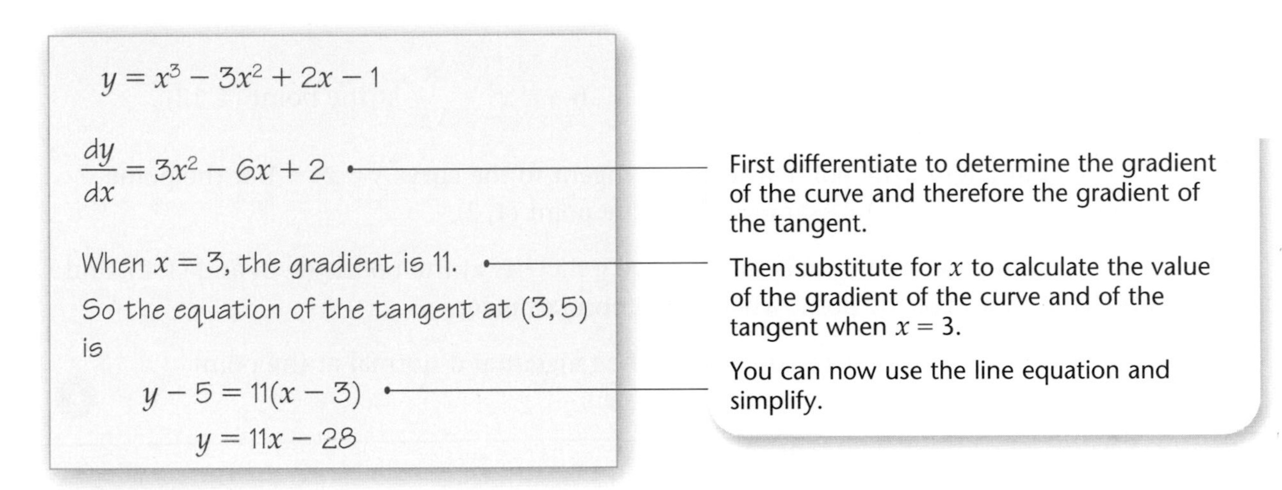

Example 14

Find the equation of the normal to the curve with equation $y = 8 - 3\sqrt{x}$ at the point where $x = 4$.

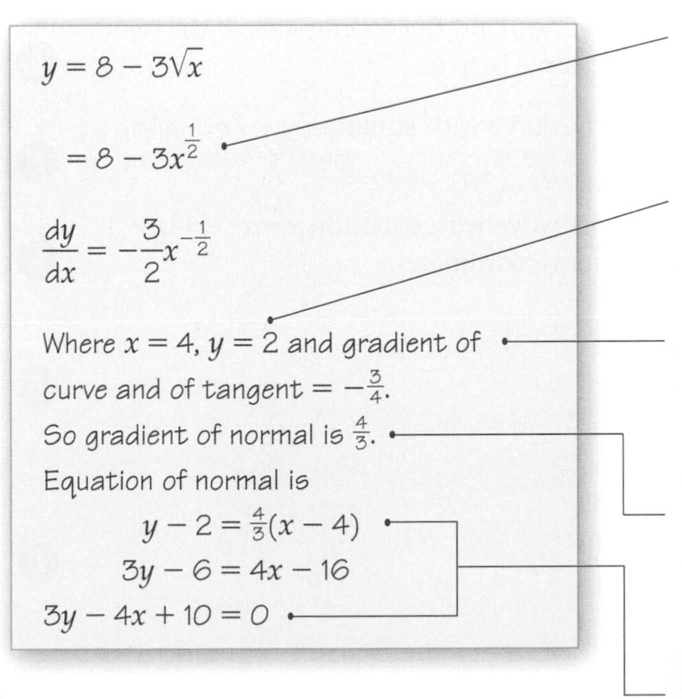

Express the function simply as powers of x, and differentiate to obtain the gradient function.

You find the y-coordinate when $x = 4$ by substituting into the equation of the curve and calculating $8 - 3\sqrt{4}$.

Then find the gradient of the curve, by calculating

$\dfrac{dy}{dx} = -\dfrac{3}{2}(4)^{-\frac{1}{2}} = -\dfrac{3}{2 \times 2}$.

Use normal gradient

$= -\dfrac{1}{\text{gradient of curve}} = -\dfrac{1}{-\frac{3}{4}} = +\dfrac{4}{3}$.

Then simplify by multiplying both sides by 3 and collecting terms.

Exercise 7H

1 Find the equation of the tangent to the curve:

a $y = x^2 - 7x + 10$ at the point (2, 0)

b $y = x + \frac{1}{x}$ at the point $(2, 2\frac{1}{2})$

c $y = 4\sqrt{x}$ at the point (9, 12)

d $y = \frac{2x - 1}{x}$ at the point (1, 1)

e $y = 2x^3 + 6x + 10$ at the point (−1, 2)

f $y = x^2 + \frac{-7}{x^2}$ at the point (1, −6)

2 Find the equation of the normal to the curves:

a $y = x^2 - 5x$ at the point (6, 6)

b $y = x^2 - \frac{8}{\sqrt{x}}$ at the point (4, 12)

3 Find the coordinates of the point where the tangent to the curve $y = x^2 + 1$ at the point (2, 5) meets the normal to the same curve at the point (1, 2).

4 Find the equations of the normals to the curve $y = x + x^3$ at the points (0, 0) and (1, 2), and find the coordinates of the point where these normals meet.

5 For $f(x) = 12 - 4x + 2x^2$, find an equation of the tangent and normal at the point where $x = -1$ on the curve with equation $y = f(x)$. **E**

Mixed exercise 7I

1 A curve is given by the equation $y = 3x^2 + 3 + \frac{1}{x^2}$, where $x > 0$. At the points A, B and C on the curve, $x = 1$, 2 and 3 respectively. Find the gradients at A, B and C. **E**

2 Taking $f(x) = \frac{1}{4}x^4 - 4x^2 + 25$, find the values of x for which $f'(x) = 0$. **E**

3 A curve is drawn with equation $y = 3 + 5x + x^2 - x^3$. Find the coordinates of the two points on the curve where the gradient of the curve is zero. **E**

4 Calculate the x-coordinates of the points on the curve with equation $y = 7x^2 - x^3$ at which the gradient is equal to 16. **E**

5 Find the x-coordinates of the two points on the curve with equation $y = x^3 - 11x + 1$ where the gradient is 1. Find the corresponding y-coordinates. **E**

6 The function f is defined by $f(x) = x + \frac{9}{x}$, $x \in \mathbb{R}$, $x \neq 0$.

a Find $f'(x)$. **b** Solve $f'(x) = 0$. **E**

7 Given that

$$y = x^{\frac{3}{2}} + \frac{48}{x}, \quad x > 0,$$

find the value of x and the value of y when $\frac{dy}{dx} = 0$. **E**

8 Given that

$$y = 3x^{\frac{1}{2}} - 4x^{-\frac{1}{2}}, \quad x > 0,$$

find $\frac{dy}{dx}$. **E**

9 A curve has equation $y = 12x^{\frac{1}{2}} - x^{\frac{3}{2}}$.

a Show that $\frac{dy}{dx} = \frac{3}{2}x^{-\frac{1}{2}}(4 - x)$

b Find the coordinates of the point on the curve where the gradient is zero. **E**

10 **a** Expand $(x^{\frac{3}{2}} - 1)(x^{-\frac{1}{2}} + 1)$.

b A curve has equation $y = (x^{\frac{3}{2}} - 1)(x^{-\frac{1}{2}} + 1)$, $x > 0$. Find $\frac{dy}{dx}$.

c Use your answer to **b** to calculate the gradient of the curve at the point where $x = 4$. **E**

11 Differentiate with respect to x:

$$2x^3 + \sqrt{x} + \frac{x^2 + 2x}{x^2}$$

E

12 The volume, V cm³, of a tin of radius r cm is given by the formula $V = \pi(40r - r^2 - r^3)$. Find the positive value of r for which $\frac{dV}{dr} = 0$, and find the value of V which corresponds to this value of r. **E**

13 The total surface area of a cylinder A cm² with a fixed volume of 1000 cubic cm is given by the formula $A = 2\pi x^2 + \frac{2000}{x}$, where x cm is the radius. Show that when the rate of change of the area with respect to the radius is zero, $x^3 = \frac{500}{\pi}$. **E**

14 The curve with equation $y = ax^2 + bx + c$ passes through the point $(1, 2)$. The gradient of the curve is zero at the point $(2, 1)$. Find the values of a, b and c. **E**

15 A curve C has equation $y = x^3 - 5x^2 + 5x + 2$.

a Find $\frac{dy}{dx}$ in terms of x.

b The points P and Q lie on C. The gradient of C at both P and Q is 2. The x-coordinate of P is 3.

i Find the x-coordinate of Q.

ii Find an equation for the tangent to C at P, giving your answer in the form $y = mx + c$, where m and c are constants.

iii If this tangent intersects the coordinate axes at the points R and S, find the length of RS, giving your answer as a surd. **E**

16 Find an equation of the tangent and the normal at the point where $x = 2$ on the curve with equation $y = \frac{8}{x} - x + 3x^2$, $x > 0$. **E**

17 The normals to the curve $2y = 3x^3 - 7x^2 + 4x$, at the points $O(0, 0)$ and $A(1, 0)$, meet at the point N.

a Find the coordinates of N.

b Calculate the area of triangle OAN. **E**

Summary of key points

1 The gradient of a curve $y = f(x)$ at a specific point is equal to the gradient of the tangent to the curve at that point.

2 The gradient of the tangent at any particular point is the rate of change of y with respect to x.

3 The gradient formula for $y = f(x)$ is given by the equation gradient $= f'(x)$ where $f'(x)$ is called the derived function.

4 If $f(x) = x^n$, then $f'(x) = nx^{n-1}$.

Hint: You reduce the power by 1 and the original power multiplies the expression.

5 The gradient of a curve can also be represented by $\frac{dy}{dx}$.

6 $\frac{dy}{dx}$ is called the derivative of y with respect to x and the process of finding $\frac{dy}{dx}$ when y is given is called differentiation.

7 $y = f(x)$, $\frac{dy}{dx} = f'(x)$

8 $y = x^n$, $\frac{dy}{dx} = nx^{n-1}$ for all real values of n.

9 It can also be shown that if $y = ax^n$ where a is a constant, then $\frac{dy}{dx} = nax^{n-1}$.

Hint: You again reduce the power by 1 and the original power multiplies the expression.

10 If $y = f(x) \pm g(x)$ then $\frac{dy}{dx} = f'(x) \pm g'(x)$.

11 A second order derivative is written as $\frac{d^2y}{dx^2}$ or $f''(x)$, using function notation.

12 You find the rate of change of a function f at a particular point by using $f'(x)$ and substituting in the value of x.

13 The equation of the tangent to the curve $y = f(x)$ at point A, $(a, f(a))$ is $y - f(a) = f'(a)(x - a)$.

14 The equation of the normal to the curve $y = f(x)$ at point A, $(a, f(a))$ is $y - f(a) = -\frac{1}{f'(a)}(x - a)$.

After completing this chapter you should be able to

1 integrate simple functions

2 understand the symbol $\int dx$

3 find the constant of integration by substituting in a given point (x, y).

Integration

Did you know?

...that understanding integration will help you find the area and volume of most standard shapes.

For example, the volume of a sphere

$V = \frac{4}{3}\pi r^3$

can be proved by integration.

You will integrate more complex functions in Core 2, 3 and 4. Integration can be used to solve many real life problems from the world of Science and Economics.

8.1 You can integrate functions of the form $f(x) = ax^n$ where $n \in \mathbb{R}$ and a is a constant.

In Chapter 7 you saw that if $y = x^2$

then $\frac{dy}{dx} = 2x$.

Also if $y = x^2 + 1$

then $\frac{dy}{dx} = 2x$.

So if $y = x^2 + c$ where c is some constant

then $\frac{dy}{dx} = 2x$.

Integration is the process of finding y when you know $\frac{dy}{dx}$.

If $\frac{dy}{dx} = 2x$

then $y = x^2 + c$ where c is some constant.

■ **If $\frac{dy}{dx} = x^n$, then $y = \frac{1}{n+1}x^{n+1} + c$, $n \neq -1$.**

Hint: This is called indefinite integration because you cannot find the constant.

Example 1

Find y for the following:

a $\frac{dy}{dx} = x^4$ **b** $\frac{dy}{dx} = x^{-5}$

a $\frac{dy}{dx} = x^4$

$\frac{dy}{dx} = x^n$ where $n = 4$.

So use $y = \frac{1}{n+1}x^{n+1} + c$ for $n = 4$.

$y = \frac{x^5}{5} + c$

Raise the power by 1.

Divide by the new power and don't forget to add c.

b $\frac{dy}{dx} = x^{-5}$

$y = \frac{x^{-4}}{-4} + c$

$= -\frac{1}{4}x^{-4} + c$

Remember raising the power by 1 gives $-5 + 1 = -4$.

Divide by the new power (-4) and add c.

Example 2

Find y for the following:

a $\frac{dy}{dx} = 2x^3$ **b** $\frac{dy}{dx} = 3x^{\frac{1}{2}}$

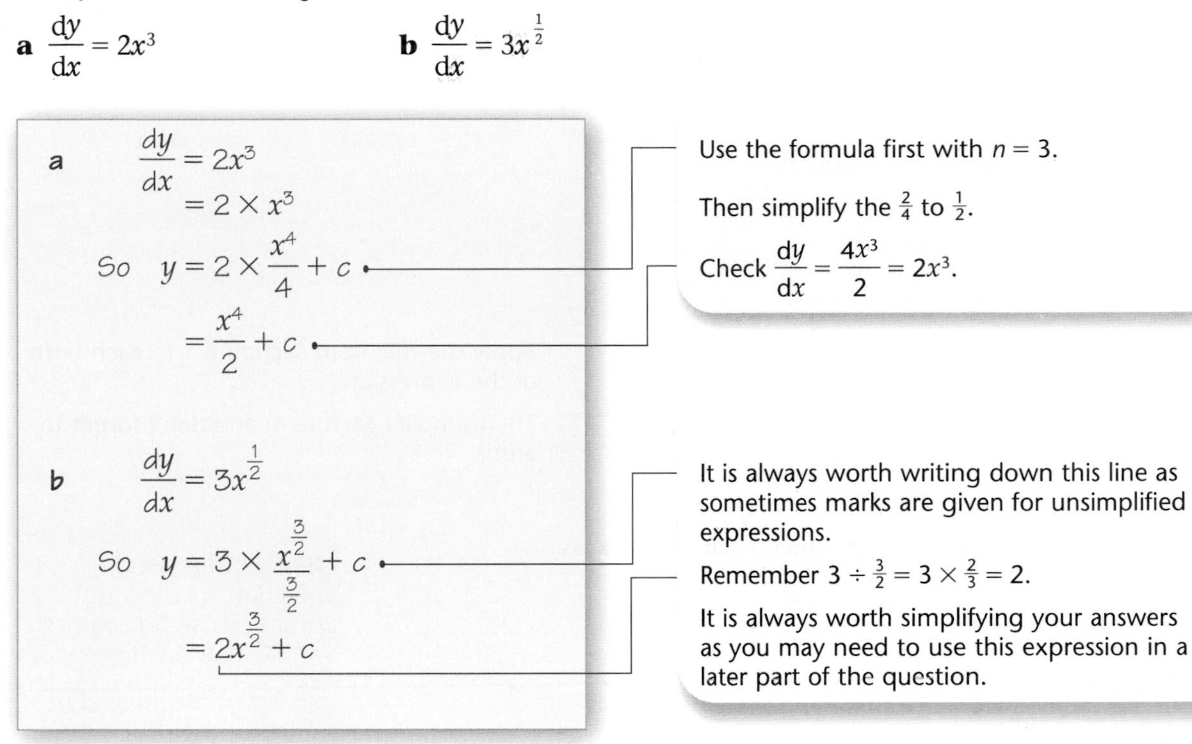

Use the formula first with $n = 3$.

Then simplify the $\frac{2}{4}$ to $\frac{1}{2}$.

Check $\frac{dy}{dx} = \frac{4x^3}{2} = 2x^3$.

It is always worth writing down this line as sometimes marks are given for unsimplified expressions.

Remember $3 \div \frac{3}{2} = 3 \times \frac{2}{3} = 2$.

It is always worth simplifying your answers as you may need to use this expression in a later part of the question.

Notice that you treat $\frac{dy}{dx} = x^n$ and $\frac{dy}{dx} = kx^n$ in the same way. You only consider the x^n term when integrating.

So in general

■ **If $\frac{dy}{dx} = kx^n$, then $y = \frac{kx^{n+1}}{n+1} + c$, $n \neq -1$.**

Exercise 8A

Find an expression for y when $\frac{dy}{dx}$ is the following:

1 x^5 **2** $10x^4$ **3** $3x^2$

4 $-x^{-2}$ **5** $-4x^{-3}$ **6** $x^{\frac{2}{3}}$

7 $4x^{\frac{1}{2}}$ **8** $-2x^6$ **9** $3x^5$

10 $3x^{-4}$ **11** $x^{-\frac{1}{2}}$ **12** $5x^{-\frac{3}{2}}$

13 $-2x^{-\frac{3}{2}}$ **14** $6x^{\frac{1}{3}}$ **15** $36x^{11}$

16 $-14x^{-8}$ **17** $-3x^{-\frac{2}{3}}$ **18** -5

19 $6x$ **20** $2x^{-0.4}$

8.2 You can apply the principle of integration separately to each term of $\frac{dy}{dx}$.

Example 3

Given $\frac{dy}{dx} = 6x + 2x^{-3} - 3x^{\frac{1}{2}}$, find y.

$$y = \frac{6x^2}{2} + \frac{2}{-2}x^{-2} - \frac{3}{\frac{3}{2}}x^{\frac{3}{2}} + c$$

$$= 3x^2 - x^{-2} - 2x^{\frac{3}{2}} + c$$

Apply the rule from Section 8.1 to each term of the expression.

Then simplify each term and don't forget to add c.

In Chapter 7 you saw that if $y = \text{f}(x)$, then $\frac{dy}{dx} = \text{f}'(x)$.

Hint: Both types of notation are used in the next exercise. Sometimes we say that the integral of $\frac{dy}{dx}$ is y or the integral of $\text{f}'(x)$ is $\text{f}(x)$.

Exercise 8B

1 Find y when $\frac{dy}{dx}$ is given by the following expressions. In each case simplify your answer:

a $4x - x^{-2} + 6x^{\frac{1}{2}}$

b $15x^2 + 6x^{-3} - 3x^{-\frac{5}{2}}$

c $x^3 - \frac{3}{2}x^{-\frac{1}{2}} - 6x^{-2}$

d $4x^3 + x^{-\frac{2}{3}} - x^{-2}$

e $4 - 12x^{-4} + 2x^{-\frac{1}{2}}$

f $5x^{\frac{2}{3}} - 10x^4 + x^{-3}$

g $-\frac{4}{3}x^{-\frac{4}{3}} - 3 + 8x$

h $5x^4 - x^{-\frac{3}{2}} - 12x^{-5}$

2 Find $\text{f}(x)$ when $\text{f}'(x)$ is given by the following expressions. In each case simplify your answer:

a $12x + \frac{3}{2}x^{-\frac{3}{2}} + 5$

b $6x^5 + 6x^{-7} - \frac{1}{6}x^{-\frac{7}{6}}$

c $\frac{1}{2}x^{-\frac{1}{2}} - \frac{1}{2}x^{-\frac{3}{2}}$

d $10x + 8x^{-3}$

e $2x^{-\frac{1}{3}} + 4x^{-\frac{5}{3}}$

f $9x^2 + 4x^{-3} + \frac{1}{4}x^{-\frac{1}{2}}$

g $x^2 + x^{-2} + x^{\frac{1}{2}}$

h $-2x^{-3} - 2x + 2x^{\frac{1}{2}}$

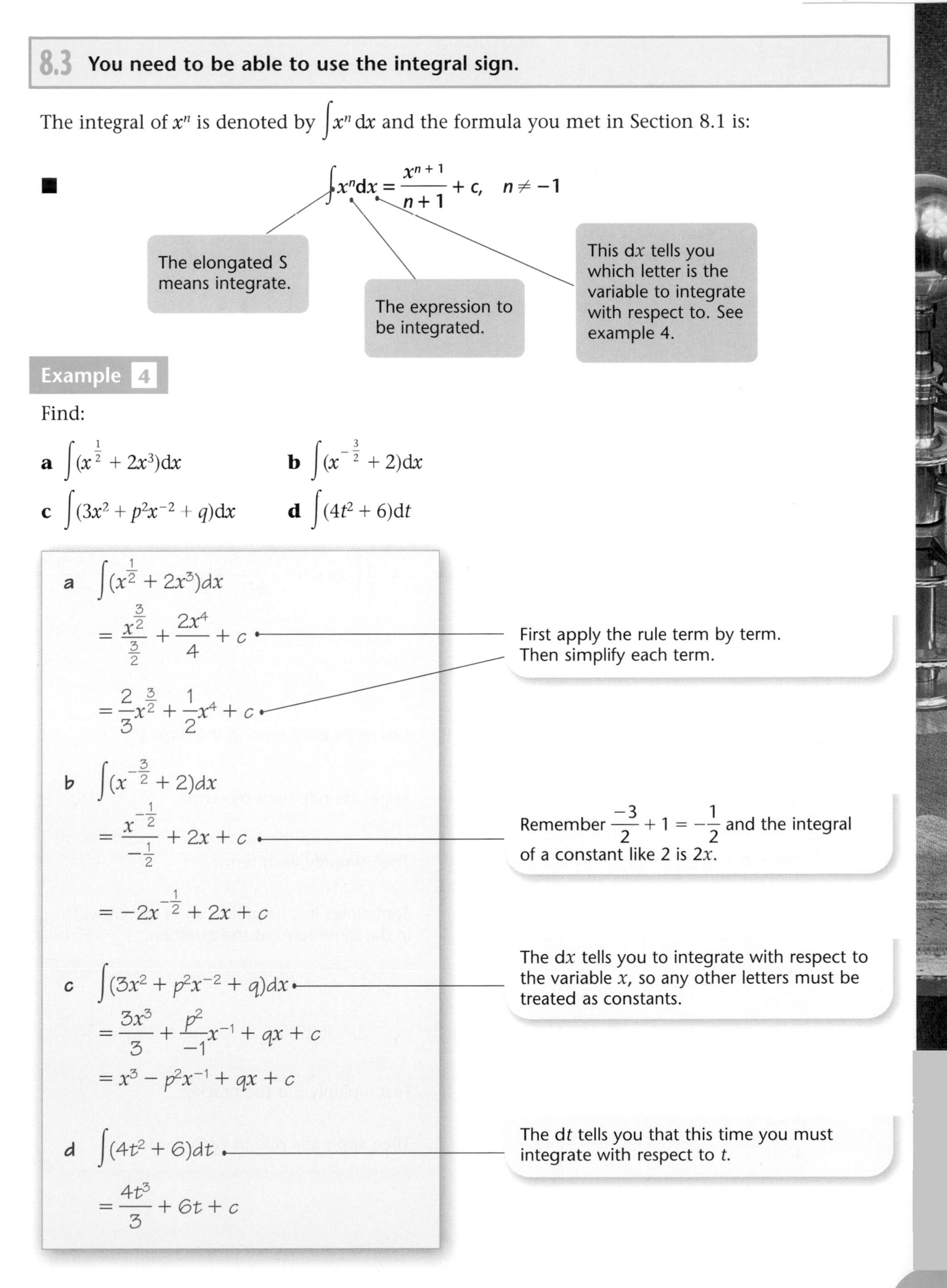

8.3 You need to be able to use the integral sign.

The integral of x^n is denoted by $\int x^n\,dx$ and the formula you met in Section 8.1 is:

■ $$\int x^n dx = \frac{x^{n+1}}{n+1} + c, \quad n \neq -1$$

The elongated S means integrate.

The expression to be integrated.

This dx tells you which letter is the variable to integrate with respect to. See example 4.

Example 4

Find:

a $\int (x^{\frac{1}{2}} + 2x^3)dx$ **b** $\int (x^{-\frac{3}{2}} + 2)dx$

c $\int (3x^2 + p^2x^{-2} + q)dx$ **d** $\int (4t^2 + 6)dt$

a $\int (x^{\frac{1}{2}} + 2x^3)dx$

$= \frac{x^{\frac{3}{2}}}{\frac{3}{2}} + \frac{2x^4}{4} + c$

First apply the rule term by term. Then simplify each term.

$= \frac{2}{3}x^{\frac{3}{2}} + \frac{1}{2}x^4 + c$

b $\int (x^{-\frac{3}{2}} + 2)dx$

$= \frac{x^{-\frac{1}{2}}}{-\frac{1}{2}} + 2x + c$

Remember $\frac{-3}{2} + 1 = -\frac{1}{2}$ and the integral of a constant like 2 is $2x$.

$= -2x^{-\frac{1}{2}} + 2x + c$

c $\int (3x^2 + p^2x^{-2} + q)dx$

The dx tells you to integrate with respect to the variable x, so any other letters must be treated as constants.

$= \frac{3x^3}{3} + \frac{p^2}{-1}x^{-1} + qx + c$

$= x^3 - p^2x^{-1} + qx + c$

d $\int (4t^2 + 6)dt$

The dt tells you that this time you must integrate with respect to t.

$= \frac{4t^3}{3} + 6t + c$

Exercise 8C

Find the following integrals.

1 $\int(x^3 + 2x)\text{d}x$

2 $\int(2x^{-2} + 3)\text{d}x$

3 $\int(5x^{\frac{3}{2}} - 3x^2)\text{d}x$

4 $\int(2x^{\frac{1}{2}} - 2x^{-\frac{1}{2}} + 4)\text{d}x$

5 $\int(4x^3 - 3x^{-4} + r)\text{d}x$

6 $\int(3t^2 - t^{-2})\text{d}t$

7 $\int(2t^2 - 3t^{-\frac{3}{2}} + 1)\text{d}t$

8 $\int(x + x^{-\frac{1}{2}} + x^{-\frac{3}{2}})\text{d}x$

9 $\int(px^4 + 2t + 3x^{-2})\text{d}x$

10 $\int(pt^3 + q^2 + px^3)\text{d}t$

8.4 **You need to simplify an expression into separate terms of the form x^n, $n \in \mathbb{R}$, before you integrate.**

Example 5

Find the following integrals:

a $\int\left(\frac{2}{x^3} - 3\sqrt{x}\right)\text{d}x$ **b** $\int x\left(x^2 + \frac{2}{x}\right)\text{d}x$ **c** $\int\left[(2x)^2 + \frac{\sqrt{x} + 5}{x^2}\right]\text{d}x$

a $\int\left(\frac{2}{x^3} - 3\sqrt{x}\right)\text{d}x$

$= \int(2x^{-3} - 3x^{\frac{1}{2}})\text{d}x$ — First write each term in the form x^n.

$= \frac{2}{-2}x^{-2} - \frac{3}{\frac{3}{2}}x^{\frac{3}{2}} + c$ — Apply the rule term by term.

$= -x^{-2} - 2x^{\frac{3}{2}} + c$ — Then simplify each term.

or $= -\frac{1}{x^2} - 2\sqrt{x^3} + c$ — Sometimes it is helpful to write the answer in the same form as the question.

b $\int x\left(x^2 + \frac{2}{x}\right)\text{d}x$

$= \int(x^3 + 2)\text{d}x$ — First multiply out the bracket.

$= \frac{x^4}{4} + 2x + c$ — Then apply the rule to each term.

c $\int\left[(2x)^2 + \frac{\sqrt{x}+5}{x^2}\right]dx$

$= \int\left[4x^2 + \frac{x^{\frac{1}{2}}}{x^2} + \frac{5}{x^2}\right]dx$ — Simplify $(2x)^2$ and write $\sqrt{x}$ as $x^{\frac{1}{2}}$.

$= \int(4x^2 + x^{-\frac{3}{2}} + 5x^{-2})dx$ — Write each term in the x^n form.

$= \frac{4}{3}x^3 + \frac{x^{-\frac{1}{2}}}{-\frac{1}{2}} + \frac{5x^{-1}}{-1} + c$ — Apply the rule term by term.

$= \frac{4}{3}x^3 - 2x^{-\frac{1}{2}} - 5x^{-1} + c$

or $= \frac{4}{3}x^3 - \frac{2}{\sqrt{x}} - \frac{5}{x} + c$ — Finally simplify the answer.

Exercise 8D

1 Find the following integrals:

a $\int(2x+3)x^2\,dx$ **b** $\int\frac{(2x^2+3)}{x^2}\,dx$ **c** $\int(2x+3)^2\,dx$

d $\int(2x+3)(x-1)\,dx$ **e** $\int(2x+3)\sqrt{x}\,dx$

2 Find $\int f(x)dx$ when $f(x)$ is given by the following:

a $(x+2)^2$ **b** $\left(x+\frac{1}{x}\right)^2$ **c** $(\sqrt{x}+2)^2$

d $\sqrt{x}(x+2)$ **e** $\left(\frac{x+2}{\sqrt{x}}\right)$ **f** $\left(\frac{1}{\sqrt{x}}+2\sqrt{x}\right)$

3 Find the following integrals:

a $\int\left(3\sqrt{x}+\frac{1}{x^2}\right)dx$ **b** $\int\left(\frac{2}{\sqrt{x}}+3x^2\right)dx$

c $\int\left(x^{\frac{2}{3}}+\frac{4}{x^3}\right)dx$ **d** $\int\left(\frac{2+x}{x^3}+3\right)dx$

e $\int(x^2+3)(x-1)dx$ **f** $\int\left(\frac{2}{\sqrt{x}}+3x\sqrt{x}\right)dx$

g $\int(x-3)^2\,dx$ **h** $\int\frac{(2x+1)^2}{\sqrt{x}}\,dx$

i $\int\left(3+\frac{\sqrt{x}+6x^3}{x}\right)dx$ **j** $\int\sqrt{x}(\sqrt{x}+3)^2\,dx$

8.5 You can find the constant of integration, c, when you are given any point (x, y) that the curve of the function passes through.

Example 6

The curve C with equation $y = \mathrm{f}(x)$ passes through the point (4, 5). Given that $\mathrm{f}'(x) = \dfrac{x^2 - 2}{\sqrt{x}}$, find the equation of C.

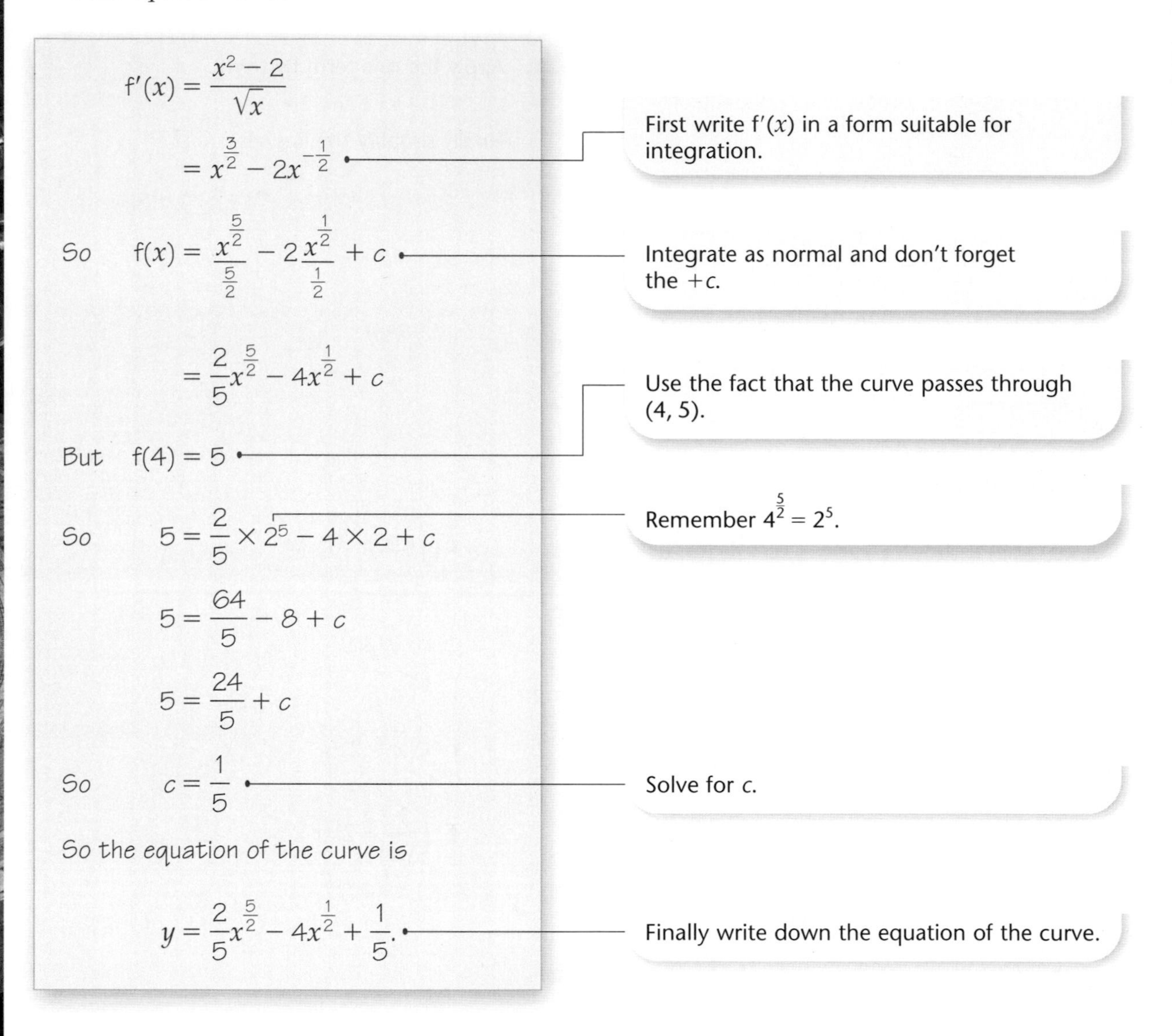

$$\mathrm{f}'(x) = \frac{x^2 - 2}{\sqrt{x}}$$

$$= x^{\frac{3}{2}} - 2x^{-\frac{1}{2}}$$

First write $\mathrm{f}'(x)$ in a form suitable for integration.

So $$\mathrm{f}(x) = \frac{x^{\frac{5}{2}}}{\frac{5}{2}} - 2\frac{x^{\frac{1}{2}}}{\frac{1}{2}} + c$$

Integrate as normal and don't forget the $+c$.

$$= \frac{2}{5}x^{\frac{5}{2}} - 4x^{\frac{1}{2}} + c$$

But $\mathrm{f}(4) = 5$

Use the fact that the curve passes through (4, 5).

So $$5 = \frac{2}{5} \times 2^5 - 4 \times 2 + c$$

Remember $4^{\frac{5}{2}} = 2^5$.

$$5 = \frac{64}{5} - 8 + c$$

$$5 = \frac{24}{5} + c$$

So $$c = \frac{1}{5}$$

Solve for c.

So the equation of the curve is

$$y = \frac{2}{5}x^{\frac{5}{2}} - 4x^{\frac{1}{2}} + \frac{1}{5}.$$

Finally write down the equation of the curve.

Exercise 8E

1 Find the equation of the curve with the given derivative of y with respect to x that passes through the given point:

a $\dfrac{\mathrm{d}y}{\mathrm{d}x} = 3x^2 + 2x;$ point (2, 10)

b $\dfrac{\mathrm{d}y}{\mathrm{d}x} = 4x^3 + \dfrac{2}{x^3} + 3;$ point (1, 4)

c $\frac{dy}{dx} = \sqrt{x} + \frac{1}{4}x^2$; point (4, 11)

d $\frac{dy}{dx} = \frac{3}{\sqrt{x}} - x$; point (4, 0)

e $\frac{dy}{dx} = (x + 2)^2$; point (1, 7)

f $\frac{dy}{dx} = \frac{x^2 + 3}{\sqrt{x}}$; point (0, 1)

2 The curve C, with equation $y = f(x)$, passes through the point (1, 2) and $f'(x) = 2x^3 - \frac{1}{x^2}$. Find the equation of C in the form $y = f(x)$.

3 The gradient of a particular curve is given by $\frac{dy}{dx} = \frac{\sqrt{x} + 3}{x^2}$. Given that the curve passes through the point (9, 0), find an equation of the curve.

4 A set of curves, that each pass through the origin, have equations $y = f_1(x)$, $y = f_2(x)$, $y = f_3(x)$ ... where $f'_n(x) = f_{n-1}(x)$ and $f_1(x) = x^2$.

a Find $f_2(x)$, $f_3(x)$.

b Suggest an expression for $f_n(x)$.

5 A set of curves, with equations $y = f_1(x)$, $y = f_2(x)$, $y = f_3(x)$ ... all pass through the point (0, 1) and they are related by the property $f'_n(x) = f_{n-1}(x)$ and $f_1(x) = 1$.
Find $f_2(x)$, $f_3(x)$, $f_4(x)$.

Mixed exercise 8F

1 Find:

a $\int (x + 1)(2x - 5)dx$ **b** $\int (x^{\frac{1}{3}} + x^{-\frac{1}{3}})dx$.

2 The gradient of a curve is given by $f'(x) = x^2 - 3x - \frac{2}{x^2}$. Given that the curve passes through the point (1, 1), find the equation of the curve in the form $y = f(x)$.

3 Find:

a $\int (8x^3 - 6x^2 + 5)dx$ **b** $\int (5x + 2)x^{\frac{1}{2}}\,dx$.

4 Given $y = \frac{(x + 1)(2x - 3)}{\sqrt{x}}$, find $\int y\,dx$.

5 Given that $\frac{dx}{dt} = 3t^2 - 2t + 1$ and that $x = 2$ when $t = 1$, find the value of x when $t = 2$.

6 Given $y = 3x^{\frac{1}{2}} + 2x^{-\frac{1}{2}}$, $x > 0$, find $\int y\,dx$.

7 Given that $\frac{dx}{dt} = (t + 1)^2$ and that $x = 0$ when $t = 2$, find the value of x when $t = 3$.

8 Given that $y^{\frac{1}{2}} = x^{\frac{1}{3}} + 3$:

a show that $y = x^{\frac{2}{3}} + Ax^{\frac{1}{3}} + B$, where A and B are constants to be found

b hence find $\int y \mathrm{d}x$.

E

9 Given that $y = 3x^{\frac{1}{2}} - 4x^{-\frac{1}{2}}$ $(x > 0)$:

a find $\frac{\mathrm{d}y}{\mathrm{d}x}$

b find $\int y \mathrm{d}x$.

E

10 Find $\int (x^{\frac{1}{2}} - 4)(x^{-\frac{1}{2}} - 1)\mathrm{d}x$.

E

Summary of key points

1 If $\frac{\mathrm{d}y}{\mathrm{d}x} = x^n$, then $y = \frac{1}{n+1}x^{n+1} + c$ $(n \neq -1)$.

2 If $\frac{\mathrm{d}y}{\mathrm{d}x} = kx^n$, then $y = \frac{kx^{n+1}}{n+1} + c$ $(n \neq -1)$.

3 $\int x^n \mathrm{d}x = \frac{x^{n+1}}{n+1} + c \quad (n \neq -1)$.

Review Exercise

1 The line L has equation $y = 5 - 2x$.

a Show that the point $P(3, -1)$ lies on L.

b Find an equation of the line, perpendicular to L, which passes through P. Give your answer in the form $ax + by + c = 0$, where a, b and c are integers. **E**

2 The points A and B have coordinates $(-2, 1)$ and $(5, 2)$ respectively.

a Find, in its simplest surd form, the length AB.

b Find an equation of the line through A and B, giving your answer in the form $ax + by + c = 0$, where a, b and c are integers.

The line through A and B meets the y-axis at the point C.

c Find the coordinates of C.

3 The line l_1 passes through the point $(9, -4)$ and has gradient $\frac{1}{3}$.

a Find an equation for l_1 in the form $ax + by + c = 0$, where a, b and c are integers.

The line l_2 passes through the origin O and has gradient -2. The lines l_1 and l_2 intersect at the point P.

b Calculate the coordinates of P.

Given that l_1 crosses the y-axis at the point C,

c calculate the exact area of ΔOCP. **E**

4

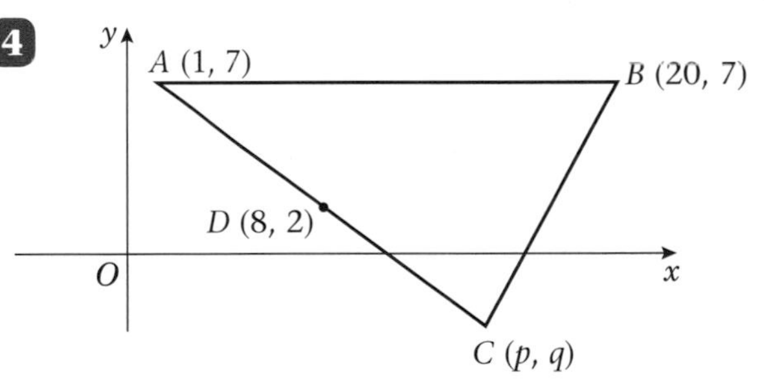

The points $A(1, 7)$, $B(20, 7)$ and $C(p, q)$ form the vertices of a triangle ABC, as shown in the figure. The point $D(8, 2)$ is the mid-point of AC.

a Find the value of p and the value of q.

The line l, which passes through D and is perpendicular to AC, intersects AB at E.

b Find an equation for l, in the form $ax + by + c = 0$, where a, b and c are integers.

c Find the exact x-coordinate of E. **E**

5 The straight line l_1 has equation $y = 3x - 6$. The straight line l_2 is perpendicular to l_1 and passes through the point (6, 2).

a Find an equation for l_2 in the form $y = mx + c$, where m and c are constants.

The lines l_1 and l_2 intersect at the point C.

b Use algebra to find the coordinates of C.

The lines l_1 and l_2 cross the x-axis at the points A and B respectively.

c Calculate the exact area of triangle ABC. **E**

6 The line l_1 has equation $6x - 4y - 5 = 0$. The line l_2 has equation $x + 2y - 3 = 0$.

a Find the coordinates of P, the point of intersection of l_1 and l_2.

The line l_1 crosses the y-axis at the point M and the line l_2 crosses the y-axis at the point N.

b Find the area of ΔMNP.

7 The 5th term of an arithmetic series is 4 and the 15th term of the series is 39.

a Find the common difference of the series.

b Find the first term of the series.

c Find the sum of the first 15 terms of the series.

8 An athlete prepares for a race by completing a practice run on each of 11 consecutive days. On each day after the first day, he runs further than he ran on the previous day. The lengths of his 11 practice runs form an arithmetic sequence with first term a km and common difference d km.

He runs 9 km on the 11th day, and he runs a total of 77 km over the 11 day period. Find the value of a and the value of d. **E**

9 The rth term of an arithmetic series is $(2r - 5)$.

a Write down the first three terms of this series.

b State the value of the common difference.

c Show that $\sum_{r=1}^{n} (2r - 5) = n(n - 4)$ **E**

10 Ahmed plans to save £250 in the year 2001, £300 in 2002, £350 in 2003, and so on until the year 2020. His planned savings form an arithmetic sequence with common difference £50.

a Find the amount he plans to save in the year 2011.

b Calculate his total planned savings over the 20 year period from 2001 to 2020.

Ben also plans to save money over the same 20 year period. He saves £A in the year 2001 and his planned yearly savings form an arithmetic sequence with common difference £60.

Given that Ben's total planned savings over the 20 year period are equal to Ahmed's total planned savings over the same period,

c calculate the value of A. **E**

11 A sequence $a_1, a_2, a_3, \ldots$ is defined by

$a_1 = 3$,

$a_{n+1} = 3a_n - 5, n \geqslant 1$.

a Find the value of a_2 and the value of a_3.

b Calculate the value of $\sum_{r=1}^{5} a_r$. **E**

12 A sequence $a_1, a_2, a_3, \ldots$ is defined by

$a_1 = k$,

$a_{n+1} = 3a_n + 5, n \geqslant 1$

where k is a positive integer.

a Write down an expression for a_2 in terms of k.

b Show that $a_3 = 9k + 20$.

c **i** Find $\sum_{r=1}^{4} a_r$ in terms of k.

ii Show that $\sum_{r=1}^{4} a_r$ is divisible by 10. **E**

13 A sequence $a_1, a_2, a_3, \ldots$ is defined by

$$a_1 = k$$

$$a_{n+1} = 2a_n - 3, \qquad n \geqslant 1$$

a Show that $a_5 = 16k - 45$.

Given that $a_5 = 19$, find the value of

b k

c $\sum_{r=1}^{6} a_r$.

14 An arithmetic sequence has first term a and common difference d.

a Prove that the sum of the first n terms of the series are $\frac{1}{2}n[2a + (n - 1)d]$

Sean repays a loan over a period of n months. His monthly repayments form an arithmetic sequence.

He repays £149 in the first month, £147 in the second month, £145 in the third month, and so on. He makes his final repayment in the nth month, where $n > 21$.

b Find the amount Sean repays in the 21st month.

Over the n months, he repays a total of £5000.

c Form an equation in n, and show that your equation may be written as

$$n^2 - 150n + 5000 = 0$$

d Solve the equation in part **c**.

e State, with a reason, which of the solutions to the equation in part **c** is **not** a sensible solution to the repayment problem. *E*

15 A sequence is given by

$a_1 = 2$

$a_{n+1} = a_n^2 - ka_n, \quad n \geq 1,$

where k is a constant.

a Show that $a_3 = 6k^2 - 20k + 16$

Given that $a_3 = 2$,

b find the possible values of k.

For the larger of the possible values of k, find the value of:

c a_2

d a_5

e a_{100}.

16 Given that $y = 4x^3 - 1 + 2x^{1/2}, x > 0$, find $\frac{dy}{dx}$. *E*

17 Given that $y = 2x^2 - 6/x^3, x \neq 0$,

a find $\frac{dy}{dx}$,

b find $\int y\,dx$. *E*

18 Given that $y = 3x^2 + 4\sqrt{x}, x > 0$, find

a $\frac{dy}{dx}$,

b $\frac{d^2y}{dx^2}$,

c $\int y\,dx$. *E*

19 **a** Given that $y = 5x^3 + 7x + 3$, find

i $\frac{dy}{dx}$,

ii $\frac{d^2y}{dx^2}$,

b Find $\int (1 + 3\sqrt{x} - 1/x^2)dx$. *E*

20 The curve C has equation $y = 4x + 3x^{\frac{3}{2}} - 2x^2, x > 0$.

a Find an expression for $\frac{dy}{dx}$.

b Show that the point $P(4, 8)$ lies on C.

c Show that an equation of the normal to C at point P is $3y = x + 20$.

The normal to C at P cuts the x-axis at point Q.

d Find the length PQ, giving your answer in a simplified surd form. *E*

21 The curve C has equation $y = 4x^2 + \frac{5 - x}{x}$, $x \neq 0$. The point P on C has x-coordinate 1.

a Show that the value of $\frac{dy}{dx}$ at P is 3.

b Find an equation of the tangent to C at P.

This tangent meets the x-axis at the point $(k, 0)$.

c Find the value of k. *E*

22 The curve C has equation
$y = \frac{1}{3}x^3 - 4x^2 + 8x + 3$.
The point P has coordinates $(3, 0)$.

a Show that P lies on C.

b Find the equation of the tangent to C at P, giving your answer in the form $y = mx + c$, where m and c are constants.

Another point Q also lies on C. The tangent to C at Q is parallel to the tangent to C at P.

c Find the coordinates of Q. **E**

23 $f(x) = \dfrac{(2x + 1)(x + 4)}{\sqrt{x}}, \quad x > 0.$

a Show that $f(x)$ can be written in the form $Px^{\frac{3}{2}} + Qx^{\frac{1}{2}} + Rx^{-\frac{1}{2}}$, stating the values of the constants P, Q and R.

b Find $f'(x)$.

c Show that the tangent to the curve with equation $y = f(x)$ at the point where $x = 1$ is parallel to the line with equation $2y = 11x + 3$. **E**

24 The curve C with equation $y = f(x)$ passes through the point $(3, 5)$.
Given that $f'(x) = x^2 + 4x - 3$, find $f(x)$.

25 The curve with equation $y = f(x)$ passes through the point $(1, 6)$. Given that

$$f'(x) = 3 + (5x^2 + 2)/x^{1/2}, x > 0,$$

find $f(x)$ and simplify your answer. **E**

26 For the curve C with equation $y = f(x)$,

$$\frac{dy}{dx} = x^3 + 2x - 7$$

a find $\dfrac{d^2y}{dx^2}$

b show that $\dfrac{d^2y}{dx^2} \geqslant 2$ for all values of x.

Given that the point $P(2, 4)$ lies on C,

c find y in terms of x

d find an equation for the normal to C at P in the form $ax + by + c = 0$, where a, b and c are integers. **E**

27 For the curve C with equation $y = f(x)$,
$$\frac{dy}{dx} = \frac{1 - x^2}{x^4}$$
Given that C passes through the point $\left(\frac{1}{2}, \frac{2}{3}\right)$,

a find y in terms of x

b find the coordinates of the points on C at which $\dfrac{dy}{dx} = 0$.

28 The curve C with equation $y = f(x)$ passes through the point $(5, 65)$.
Given that $f'(x) = 6x^2 - 10x - 12$,

a use integration to find $f(x)$

b hence show that $f(x) = x(2x + 3)(x - 4)$

c sketch C, showing the coordinates of the points where C crosses the x-axis. **E**

29 The curve C has equation
$y = x^2(x - 6) + \dfrac{4}{x}, \quad x > 0.$

The points P and Q lie on C and have x-coordinates 1 and 2 respectively.

a Show that the length of PQ is $\sqrt{170}$.

b Show that the tangents to C at P and Q are parallel.

c Find an equation for the normal to C at P, giving your answer in the form $ax + by + c = 0$, where a, b and c are integers. **E**

30 **a** Factorise completely $x^3 - 7x^2 + 12x$.

b Sketch the graph of $y = x^3 - 7x^2 + 12x$, showing the coordinates of the points at which the graph crosses the x-axis.

The graph of $y = x^3 - 7x^2 + 12x$ crosses the positive x-axis at the points A and B.

The tangents to the graph at A and B meet at the point P.

c Find the coordinates of P.

Practice paper

You may not use a calculator when answering this paper.

You must show sufficient working to make your methods clear.

Answers without working may gain no credit.

1 **a** Write down the value of $16^{\frac{1}{2}}$. (1)

b Hence find the value of $16^{\frac{3}{2}}$. (2)

2 Find $\int(6x^2 + \sqrt{x})\mathrm{d}x$. (4)

3 A sequence $a_1, a_2, a_3, \ldots a_n$ is defined by

$$a_1 = 2,\ a_{n+1} = 2a_n - 1.$$

a Write down the value of a_2 and the value of a_3. (2)

b Calculate $\sum_{r=1}^{5} a_r$. (2)

4 **a** Express $(5 + \sqrt{2})^2$ in the form $a + b\sqrt{2}$, where a and b are integers. (3)

b Hence, or otherwise, simplify $(5 + \sqrt{2})^2 - (5 - \sqrt{2})^2$. (2)

5 Solve the simultaneous equations:

$$x - 3y = 6$$
$$3xy + x = 24$$ (7)

6 The points A and B have coordinates $(-3, 8)$ and $(5, 4)$ respectively.
The straight line l_1 passes through A and B.

a Find an equation for l_1, giving your answer in the form $ax + by + c = 0$, where a, b and c are integers. (4)

b Another straight line l_2 is perpendicular to l_1 and passes through the origin. Find an equation for l_2. (2)

c The lines l_1 and l_2 intersect at the point P. Use algebra to find the coordinates of P. (3)

7 On separate diagrams, sketch the curves with equations:

a $y = \frac{2}{x}$, $\quad -2 \leqslant x \leqslant 2, x \neq 0$ (2)

b $y = \frac{2}{x} - 4$, $\quad -2 \leqslant x \leqslant 2, x \neq 0$ (3)

c $y = \frac{2}{x+1}$, $-2 \leqslant x \leqslant 2$, $x \neq -1$ (3)

In each part, show clearly the coordinates of any point at which the curve meets the x-axis or the y-axis.

8 In the year 2007, a car dealer sold 400 new cars. A model for future sales assumes that sales will increase by x cars per year for the next 10 years, so that $(400 + x)$ cars are sold in 2008, $(400 + 2x)$ cars are sold in 2009, and so on.
Using this model with $x = 30$, calculate:

a The number of cars sold in the year 2016. (2)

b The total number of cars sold over the 10 years from 2007 to 2016. (3)

The dealer wants to sell at least 6000 cars over the 10-year period.
Using the same model:

c Find the least value of x required to achieve this target. (4)

9 **a** Given that

$$x^2 + 4x + c = (x + a)^2 + b$$

where a, b and c are constants:

i Find the value of a. (1)

ii Find b in terms of c. (2)

Given also that the equation $x^2 + 4x + c = 0$ has unequal real roots:

iii Find the range of possible values of c. (2)

b Find the set of values of x for which:

i $3x < 20 - x$, (2)

ii $x^2 + 4x - 21 > 0$, (4)

iii both $3x < 20 - x$ and $x^2 + 4x - 21 > 0$. (2)

10 **a** Show that $\frac{(3x-4)^2}{x^2}$ may be written as $P + \frac{Q}{x} + \frac{R}{x^2}$, where P, Q and R are constants to be found. (3)

b The curve C has equation $y = \frac{(3x-4)^2}{x^2}$, $x \neq 0$. Find the gradient of the tangent to C at the point on C where $x = -2$. (5)

c Find the equation of the normal to C at the point on C where $x = -2$, giving your answer in the form $ax + by + c = 0$, where a, b and c are integers. (5)

Examination style paper

1 Write in the form $k\sqrt{3}$, stating the value of k in each case.

a $\sqrt{75}$ (1)

b $\sqrt{12} + \sqrt{147} - \sqrt{27}$ (2)

2 a Find the value of $27^{\frac{2}{3}}$. (2)

b Simplify $\dfrac{16x^{\frac{2}{3}}}{2x}$. (2)

3

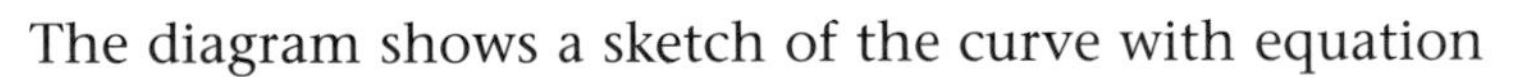

The diagram shows a sketch of the curve with equation

$$y = -\frac{6}{x}, \quad x \neq 0.$$

a On a separate diagram sketch the curve with equation

$$y = 2 - \frac{6}{x}, \quad x \neq 0.$$

showing clearly the coordinates of any point where the curve crosses the coordinate axes. (3)

b Write down the equations of the asymptotes to the curve with equation

$$y = 2 - \frac{6}{x}, \quad x \neq 0.$$ (2)

4 An arithmetic series has 1st term 49 and 15th term 7.

a Find the value of the common difference. (3)

b Find the value of the sum of the first 15 terms of the series. (3)

5 The equation $kx^2 + kx + 3 - k = 0$, where k is a constant, has no real roots.

a Show that $5k^2 - 12k < 0$. (2)

b Find the set of possible values of k. (4)

6 A sequence $a_1, a_2, a_3, \ldots$ is defined by

$$a_1 = 2$$
$$a_{n+1} = 7 - 3a_n,\ n \geqslant 1$$

a Find a_2 and a_3. (2)

b Find $\sum_{r=1}^{5} a_r$ and show that this sum is divisible by 12. (4)

7 Given that $y = 8x^3 + \dfrac{3}{\sqrt{x}} + 5,\ x > 0$

find

a $\dfrac{dy}{dx}$ (3)

b $\dfrac{d^2y}{dx^2}$ (2)

c $\int y \, dx$ (3)

8 The line l_1 has equation $2y = x - 3$ and the line l_2 has equation $5y + 2x - 18 = 0$.

a Find the gradient of l_2. (2)

The point of intersection of l_1 and l_2 is P.

b Find the coordinates of P. (3)

The lines l_1 and l_2 cross the x-axis as the points A and B respectively.

c Find the area of triangle APB. (4)

9 The curve C with equation $y = f(x)$ passes through the point (2, 4) and $f'(x) = 3(x - 1)(x + 1)$.

a Use integration to find $f(x)$. (5)

b Show that $(x - 1)^2(x + 2) = f(x)$ (3)

c Sketch C, showing the coordinates of the point where C crosses the x-axis. (3)

10 The curve C has equation

$$y = 8x + x^2 + \frac{9}{x}, \quad x \neq 0$$

The points P and Q lie on C and have x-coordinates -3 and 1 respectively.

a Find an equation of the chord PQ. (6)

b Show that the tangents to C at the points P and Q are parallel. (4)

The tangent to C at P and the normal to C at Q intersect at the point $R(17, 2)$.

c Show that $PR = 20\sqrt{2}$ (2)

d Find QR. (2)

e Explain why angle PRQ is a right angle and find the area of triangle PQR. (3)

Formulae you need to remember

These are the formulae that you need to remember for your exams. They will not be included in formulae booklets.

Quadratic equations

$ax^2 + bx + c = 0$ has roots $\dfrac{-b \pm \sqrt{b^2 - 4ac}}{2a}$

Differentiation

The derivative of x^n is nx^{n-1}

Integration

The integral of x^n is $\dfrac{1}{n+1}x^{n+1} + c,\ n \neq -1$

List of symbols and notation

The following notation will be used in all Edexcel mathematics examinations:

$\in$	is an element of
$\notin$	is not an element of
$\{x_1, x_2, \ldots\}$	the set with elements $x_1, x_2, \ldots$
$\{x: \ldots\}$	the set of all x such that ...
$\mathrm{n}(A)$	the number of elements in set A
$\varnothing$	the empty set
ξ	the universal set
A'	the complement of the set A
$\mathbb{N}$	the set of natural numbers, $\{1, 2, 3, \ldots\}$
$\mathbb{Z}$	the set of integers, $\{0, \pm 1, \pm 2, \pm 3, \ldots\}$
$\mathbb{Z}^+$	the set of positive integers, $\{1, 2, 3, \ldots\}$
$\mathbb{Z}_n$	the set of integers modulo n, $\{1, 2, 3, \ldots, n-1\}$
$\mathbb{Q}$	the set of rational numbers, $\left\{\frac{p}{q}: p \in \mathbb{Z}_u, q \in \mathbb{Z}^+\right\}$
$\mathbb{Q}^+$	the set of positive rational numbers, $\{x \in \mathbb{Q}: x > 0\}$
$\mathbb{Q}_0{}^+$	the set of positive rational numbers and zero, $\{x \in \mathbb{Q}: x \geqslant 0\}$
$\mathbb{R}$	the set of real numbers
$\mathbb{R}^+$	the set of positive real numbers, $\{x \in \mathbb{R}: x > 0\}$
$\mathbb{R}_0{}^+$	the set of positive real numbers and zero, $\{x \in \mathbb{R}: x \geqslant 0\}$
$\mathbb{C}$	the set complex numbers
(x, y)	the ordered pair x, y
$A \times B$	the cartesian products of sets A and B, ie $A \times B = \{(a, b): a \in A, b \in B\}$
$\subseteq$	is a subset of
$\subset$	is a proper subset of
$\cup$	union
$\cap$	intersection
$[a, b]$	the closed interval, $\{x \in \mathbb{R}: a \leqslant x \leqslant b\}$
$[a, b), [a, b[$	the interval, $\{x \in \mathbb{R}: a \leqslant x < b\}$
$(a, b],]a, b]$	the interval, $\{x \in \mathbb{R}: a < x \leqslant b\}$
$(a, b),]a, b[$	the open interval, $\{x \in \mathbb{R}: a < x < b\}$
$y\,R\,x$	y is related to x by the relation R
$y \sim x$	y is equivalent to x, in the context of some equivalence relation
$=$	is equal to
$\neq$	is not equal to
$\equiv$	is identical to or is congruent to

$\approx$	is approximately equal to
$\cong$	is isomorphic to
$\propto$	is proportional to
$<$	is less than
$\leqslant, \ngtr$	is less than or equal to, is not greater than
$>$	is greater than
$\geqslant, \nless$	is greater than or equal to, is not less than
∞	infinity
$p \wedge q$	p and q
$p \vee q$	p or q (or both)
$\sim p$	not p
$p \Rightarrow q$	p implies q (if p then q)
$p \Leftarrow q$	p is implied by q (if q then p)
$p \Leftrightarrow q$	p implies and is implied by q (p is equivalent to q)
$\exists$	there exists
$\forall$	for all
$a + b$	a plus b
$a - b$	a minus b
$a \times b$, ab, $a.b$	a multiplied by b
$a \div b$, $\frac{a}{b}$, a/b	a divided by b
$\sum_{i=1}^{n}$	$a_1 + a_2 + \ldots + a_n$
$\prod_{i=1}^{n}$	$a_1 \times a_2 \times \ldots \times a_n$
$\sqrt{a}$	the positive square root of a
$\lvert a \rvert$	the modulus of a
$n!$	n factorial
$\binom{n}{r}$	the binomial coefficient $\frac{n!}{r!(n-r)!}$ for $n \in \mathbb{Z}^+$
	$\frac{n(n-1)\ldots(n-r+1)}{r!}$ for $n \in \mathbb{Q}$
$f(x)$	the value of the function f at x
$f: A \rightarrow B$	f is a function under which each element of set A has an image in set B
$f: x \rightarrow y$	the function f maps the element x to the element y
f^{-1}	the inverse function of the function f
$g_o f$, gf	the composite function of f and g which is defined by $(g_o f)(x)$ or $gf(x) = g(f(x))$
$\lim_{x \to a} f(x)$	the limit of(x) of as x tends to a
Δx, δx	an increment of x
$\frac{dy}{dx}$	the derivative of y with respect to x
$\frac{d^n y}{dx^n}$	the nth derivative of y with respect to x
$f'(x), f''(x), \ldots, f^{(n)}(x)$	the first, second, …, nth derivatives of $f(x)$ with respect to x
$\int y \, dx$	the indefinite integral of y with respect to x

$\int_b^a y \, dx$	the definite integral of y with respect to x between the limits
$\frac{\partial V}{\partial x}$	the partial derivative of V with respect to x
$\dot{x}, \ddot{x}, \ldots$	the first, second, ... derivatives of x with respect to t
e	base of natural logarithms
e^x, $\exp x$	exponential function of x
$\log_a x$	logarithm to the base a of x
$\ln x$, $\log_e x$	natural logarithm of x
$\log x$, $\log_{10} x$	logarithm of x to base 10
sin, cos, tan, cosec, sec, cot	the circular functions
arcsin, arccos, arctan, arccosec, arcsec, arccot	the inverse circular functions
sinh, cosh, tanh, cosech, sech, coth	the hyperbolic functions
arsinh, arcosh, artanh, arcosech, arsech, arcoth	the inverse hyperbolic functions
i, j	square root of -1
z	a complex number, $z = x + \mathrm{i}y$
Re z	the real part of z, Re $z = x$
Im z	the imaginary part of z, Im $z = y$
$\lvert z \rvert$	the modulus of z, $\lvert z \rvert = \sqrt{(x^2 + y^2)}$
arg z	the argument of z, arg $z = \theta$, $-\pi < \theta \leqslant \pi$
z^*	the complex conjugate of z, $x - \mathrm{i}y$
$\mathbf{M}$	a matrix $\mathbf{M}$
$\mathbf{M}^{-1}$	the inverse of the matrix $\mathbf{M}$
$\mathbf{M}^{\mathrm{T}}$	the transpose of the matrix $\mathbf{M}$
det $\mathbf{M}$ or $\lvert \mathbf{M} \rvert$	the determinant of the square matrix $\mathbf{M}$
$\mathbf{a}$	the vector $\mathbf{a}$
$\overrightarrow{AB}$	the vector represented in magnitude and direction by the directed line segment AB
$\hat{\mathbf{a}}$	a unit vector in the direction of $\mathbf{a}$
$\mathbf{i}, \mathbf{j}, \mathbf{k}$	unit vectors in the direction of the cartesian coordinate axes
$\lvert \mathbf{a} \rvert$, a	the magnitude of $\mathbf{a}$
$\lvert \overrightarrow{AB} \rvert$	the magnitude of $\overrightarrow{AB}$
$\mathbf{a} . \mathbf{b}$	the scalar product of $\mathbf{a}$ and $\mathbf{b}$
$\mathbf{a} \times \mathbf{b}$	the vector product of $\mathbf{a}$ and $\mathbf{b}$

Answers

Exercise 1A

1 $7x + y$
2 $10t - 2r$
3 $8m + n - 7p$
4 $3a + 2ac - 4ab$
5 $6x^2$
6 $2m^2n + 3mn^2$
7 $2x^2 + 6x + 8$
8 $9x^2 - 2x - 1$
9 $6x^2 - 12x - 10$
10 $10c^2d + 8cd^2$
11 $8x^2 + 3x + 13$
12 $a^2b - 2a$
13 $3x^2 + 14x + 19$
14 $8x^2 - 9x + 13$
15 $a + 4b + 14c$
16 $9d^2 - 2c$
17 $20 - 6x$
18 $13 - r^2$

Exercise 1B

1 x^7
2 $6x^5$
3 $2p^2$
4 $3x^{-2}$
5 k^5
6 y^{10}
7 $5x^8$
8 p^2
9 $2a^3$
10 $2p^{-7}$
11 $6a^{-9}$
12 $3a^2b^{-2}$
13 $27x^8$
14 $24x^{11}$
15 $63a^{12}$
16 $32y^6$
17 $4a^6$
18 $6a^{12}$

Exercise 1C

1 $9x - 18$
2 $x^2 + 9x$
3 $-12y + 9y^2$
4 $xy + 5x$
5 $-3x^2 - 5x$
6 $-20x^2 - 5x$
7 $4x^2 + 5x$
8 $-15y + 6y^3$
9 $-10x^2 + 8x$
10 $3x^3 - 5x^2$
11 $4x - 1$
12 $2x - 4$
13 $3x^3 - 2x^2 + 5x$
14 $14y^2 - 35y^3 + 21y^4$
15 $-10y^2 + 14y^3 - 6y^4$
16 $4x + 10$
17 $11x - 6$
18 $7x^2 - 3x + 7$
19 $-2x^2 + 26x$
20 $-9x^3 + 23x^2$

Exercise 1D

1 $4(x + 2)$
2 $6(x - 4)$
3 $5(4x + 3)$
4 $2(x^2 + 2)$
5 $4(x^2 + 5)$
6 $6x(x - 3)$
7 $x(x - 7)$
8 $2x(x + 2)$
9 $x(3x - 1)$
10 $2x(3x - 1)$
11 $5y(2y - 1)$
12 $7x(5x - 4)$
13 $x(x + 2)$
14 $y(3y + 2)$
15 $4x(x + 3)$
16 $5y(y - 4)$
17 $3xy(3y + 4x)$
18 $2ab(3 - b)$
19 $5x(x - 5y)$
20 $4xy(3x + 2y)$
21 $5y(3 - 4z^2)$
22 $6(2x^2 - 5)$
23 $xy(y - x)$
24 $4y(3y - x)$

Exercise 1E

1 $x(x + 4)$
2 $2x(x + 3)$
3 $(x + 8)(x + 3)$
4 $(x + 6)(x + 2)$
5 $(x + 8)(x - 5)$
6 $(x - 6)(x - 2)$
7 $(x + 2)(x + 3)$
8 $(x - 6)(x + 4)$
9 $(x - 5)(x + 2)$
10 $(x + 5)(x - 4)$
11 $(2x + 1)(x + 2)$
12 $(3x - 2)(x + 4)$
13 $(5x - 1)(x - 3)$
14 $2(3x + 2)(x - 2)$
15 $(2x - 3)(x + 5)$
16 $2(x^2 + 3)(x^2 + 4)$
17 $(x + 2)(x - 2)$
18 $(x + 7)(x - 7)$
19 $(2x + 5)(2x - 5)$
20 $(3x + 5y)(3x - 5y)$
21 $4(3x + 1)(3x - 1)$
22 $2(x + 5)(x - 5)$
23 $2(3x - 2)(x - 1)$
24 $3(5x - 1)(x + 3)$

Exercise 1F

1 **a** x^5 **b** x^{-2} **c** x^4
d x^3 **e** x^5 **f** $12x^0 = 12$
g $3x^{\frac{1}{2}}$ **h** $5x$ **i** $6x^{-1}$

2 **a** ± 5 **b** ± 9 **c** 3 **d** $\frac{1}{16}$
e $\pm\frac{1}{3}$ **f** $\frac{1}{-125}$ **g** 1 **h** ± 6
i $\pm\frac{125}{64}$ **j** $\frac{9}{4}$ **k** $\frac{5}{6}$ **l** $\frac{64}{49}$

Exercise 1G

1 $2\sqrt{7}$
2 $6\sqrt{2}$
3 $5\sqrt{2}$
4 $4\sqrt{2}$
5 $3\sqrt{10}$
6 $\sqrt{3}$
7 $\sqrt{3}$
8 $6\sqrt{5}$
9 $7\sqrt{2}$
10 $12\sqrt{7}$
11 $-3\sqrt{7}$
12 $9\sqrt{5}$
13 $23\sqrt{5}$
14 2
15 $19\sqrt{3}$

Exercise 1H

1 $\dfrac{\sqrt{5}}{5}$
2 $\dfrac{\sqrt{11}}{11}$
3 $\dfrac{\sqrt{2}}{2}$
4 $\dfrac{\sqrt{5}}{5}$
5 $\frac{1}{2}$
6 $\frac{1}{4}$
7 $\dfrac{\sqrt{13}}{13}$
8 $\frac{1}{3}$
9 $\dfrac{1 - \sqrt{3}}{-2}$
10 $\dfrac{2 - \sqrt{5}}{-1}$
11 $\dfrac{3 + \sqrt{7}}{2}$
12 $3 + \sqrt{5}$
13 $\dfrac{\sqrt{5} + \sqrt{3}}{2}$
14 $\dfrac{(3 - \sqrt{2})(4 + \sqrt{5})}{11}$
15 $\dfrac{5(2 - \sqrt{5})}{-1}$
16 $5(4 + \sqrt{14})$
17 $\dfrac{11(3 - \sqrt{11})}{-2}$
18 $\dfrac{5 - \sqrt{21}}{-2}$
19 $\dfrac{14 - \sqrt{187}}{3}$
20 $\dfrac{35 + \sqrt{1189}}{6}$
21 -1

Mixed exercise 1I

1 **a** y^8 **b** $6x^7$ **c** $32x$ **d** $12b^9$

2 **a** $15y + 12$ **b** $15x^2 - 25x^3 + 10x^4$
c $16x^2 + 13x$ **d** $9x^3 - 3x^2 + 4x$

3 **a** $x(3x + 4)$ **b** $2y(2y + 5)$
c $x(x + y + y^2)$ **d** $2xy(4y + 5x)$

4 **a** $(x + 1)(x + 2)$ **b** $3x(x + 2)$
c $(x - 7)(x + 5)$ **d** $(2x - 3)(x + 1)$
e $(5x + 2)(x - 3)$ **f** $(1 - x)(6 + x)$

5 **a** $3x^6$ **b** ± 2 **c** $6x^2$ **d** $\frac{1}{2}x^{-\frac{1}{3}}$

6 **a** $\frac{4}{9}$ **b** $\pm\frac{3375}{4913}$

7 **a** $\dfrac{\sqrt{7}}{7}$ **b** $4\sqrt{5}$

8 **a** $\dfrac{\sqrt{3}}{3}$ **b** $\sqrt{2} + 1$
c $-3\sqrt{3} - 6$ **d** $\dfrac{30 - \sqrt{851}}{-7}$

Exercise 2A

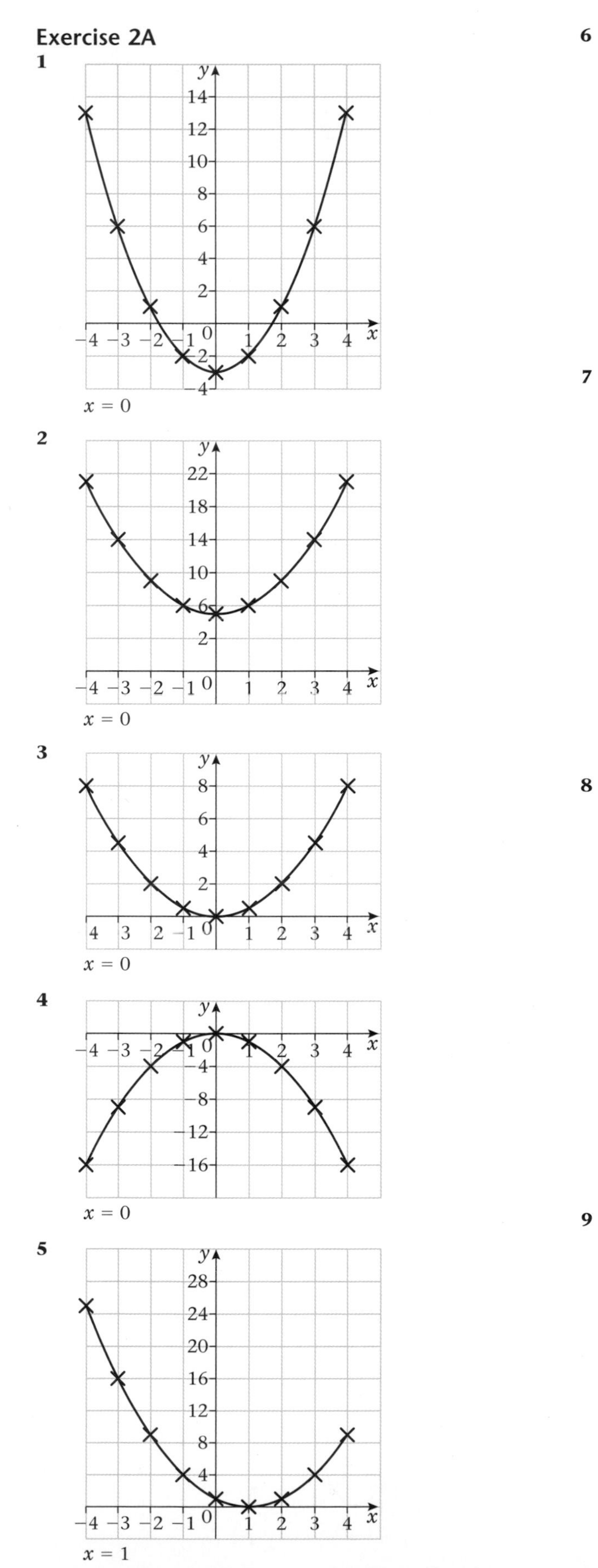

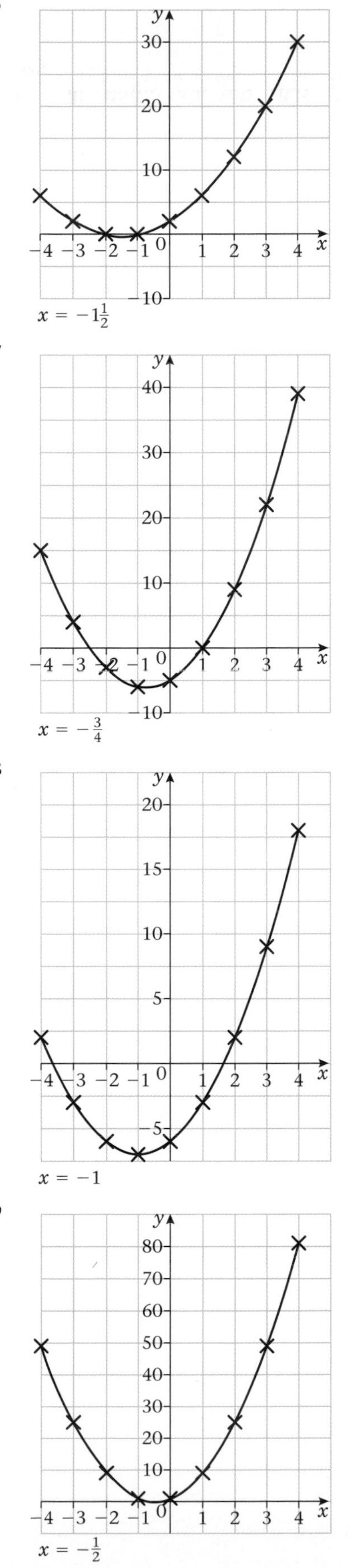

Exercise 2B

1 $x = 0$ or $x = 4$
2 $x = 0$ or $x = 25$
3 $x = 0$ or $x = 2$
4 $x = 0$ or $x = 6$
5 $x = -1$ or $x = -2$
6 $x = -1$ or $x = -4$
7 $x = -5$ or $x = -2$
8 $x = 3$ or $x = -2$
9 $x = 3$ or $x = 5$
10 $x = 4$ or $x = 5$
11 $x = 6$ or $x = -1$
12 $x = 6$ or $x = -2$
13 $x = -\frac{1}{2}$ or $x = -3$
14 $x = -\frac{1}{3}$ or $x = \frac{3}{2}$
15 $x = -\frac{2}{3}$ or $x = \frac{3}{2}$
16 $x = \frac{3}{2}$ or $x = \frac{5}{2}$
17 $x = \frac{1}{3}$ or $x = -2$
18 $x = 3$ or $x = 0$
19 $x = 13$ or $x = 1$
20 $x = 2$ or $x = -2$
21 $x = \pm\sqrt{\frac{5}{3}}$
22 $x = 3 \pm \sqrt{13}$
23 $x = \frac{1 \pm \sqrt{11}}{3}$
24 $x = 1$ or $x = -\frac{7}{6}$
25 $x = -\frac{1}{2}$ or $x = \frac{7}{3}$
26 $x = 0$ or $x = -\frac{11}{6}$

Exercise 2C

1 $(x + 2)^2 - 4$
2 $(x - 3)^2 - 9$
3 $(x - 8)^2 - 64$
4 $(x + \frac{1}{2})^2 - \frac{1}{4}$
5 $(x - 7)^2 - 49$
6 $2(x + 4)^2 - 32$
7 $3(x - 4)^2 - 48$
8 $2(x - 1)^2 - 2$
9 $5(x + 2)^2 - 20$
10 $2(x - \frac{5}{4})^2 - \frac{25}{8}$
11 $3(x + \frac{3}{2})^2 - \frac{27}{4}$
12 $3(x - \frac{1}{6})^2 - \frac{1}{12}$

Exercise 2D

1 $x = -3 \pm 2\sqrt{2}$
2 $x = -6 \pm \sqrt{33}$
3 $x = 5 \pm \sqrt{30}$
4 $x = -2 \pm \sqrt{6}$
5 $x = \frac{3}{2} \pm \frac{\sqrt{29}}{2}$
6 $x = 1 \pm \frac{3}{2}\sqrt{2}$
7 $x = \frac{1}{8} \pm \frac{\sqrt{129}}{8}$
8 No real roots
9 $x = -\frac{3}{2} \pm \frac{\sqrt{39}}{2}$
10 $x = -\frac{4}{5} \pm \frac{\sqrt{26}}{5}$

Exercise 2E

1 $\frac{-3 \pm \sqrt{5}}{2}$, -0.38 or -2.62
2 $\frac{+3 \pm \sqrt{17}}{2}$, -0.56 or 3.56
3 $-3 \pm \sqrt{3}$, -1.27 or -4.73
4 $\frac{5 \pm \sqrt{33}}{2}$, 5.37 or -0.37
5 $\frac{-5 \pm \sqrt{31}}{3}$, -3.52 or 0.19
6 $\frac{1 \pm \sqrt{2}}{2}$, 1.21 or -0.21
7 $\frac{-9 \pm \sqrt{53}}{14}$, -0.12 or -1.16
8 $\frac{-2 \pm \sqrt{19}}{5}$, 0.47 or -1.27
9 2 or $-\frac{1}{4}$
10 $\frac{-1 \pm \sqrt{78}}{11}$, 0.71 or -0.89

Exercise 2F

1 a

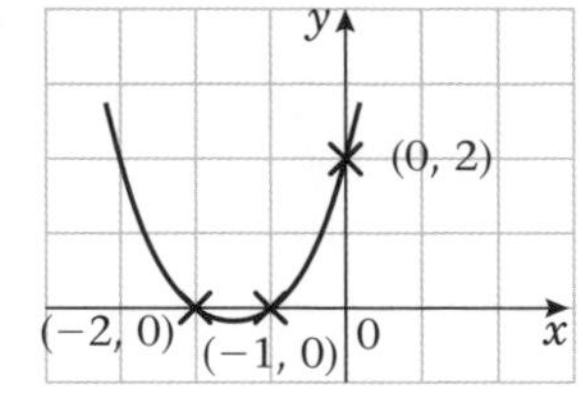

b

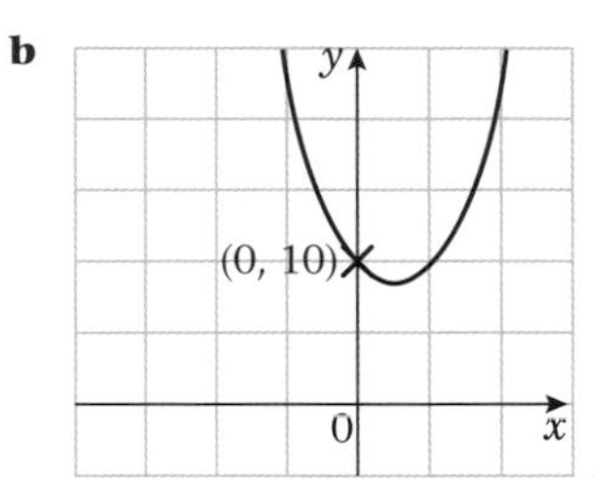

c

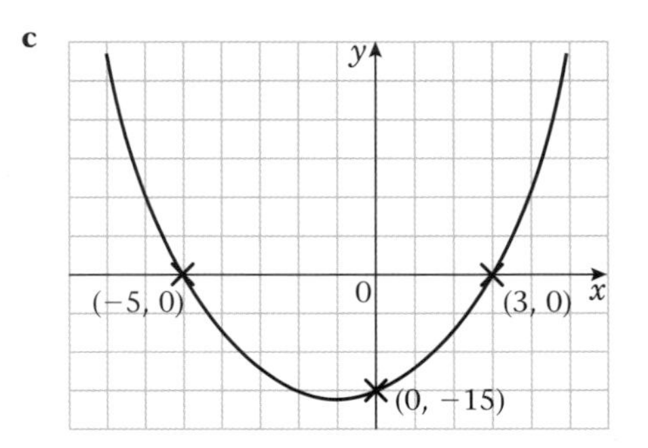

d

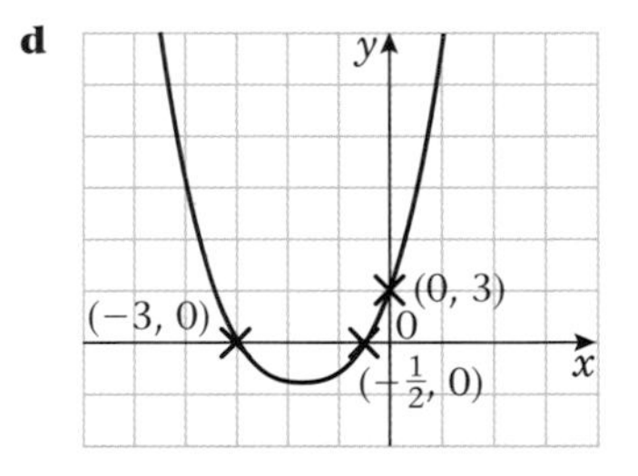

e

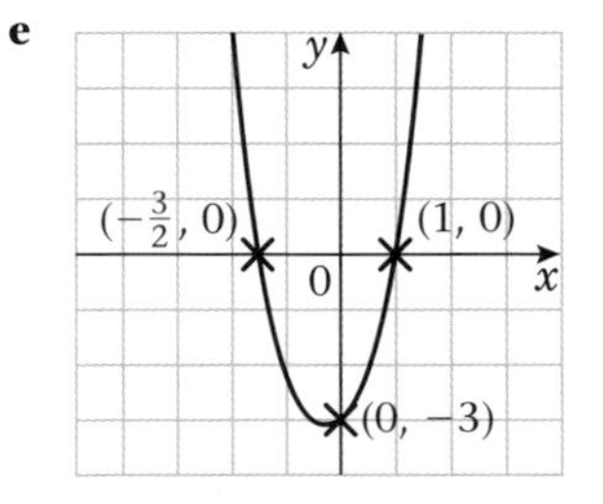

f

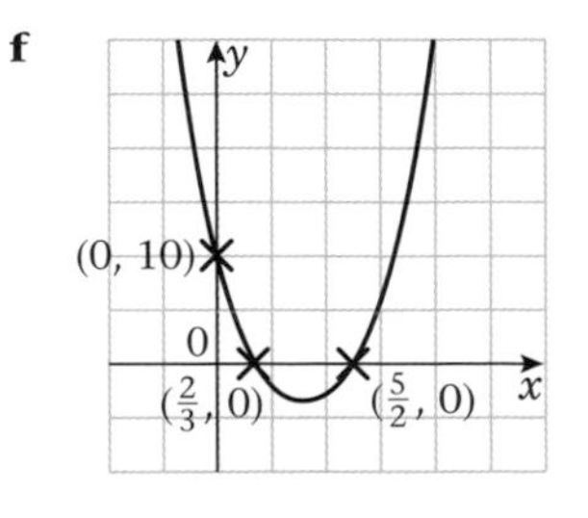

g

h

i

j

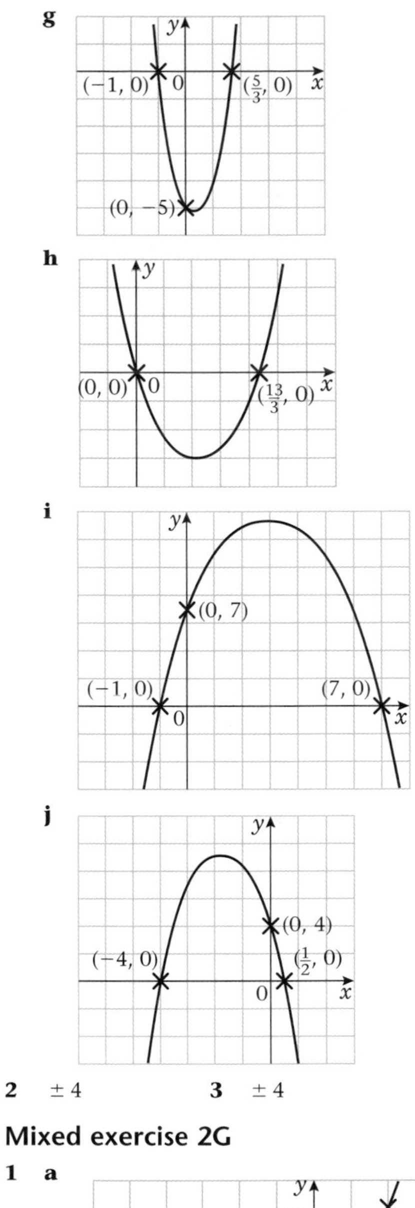

2 ± 4 **3** ± 4

Mixed exercise 2G

1 a

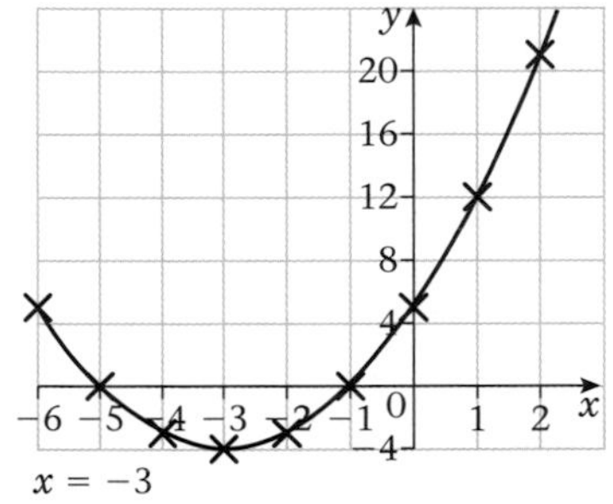

b

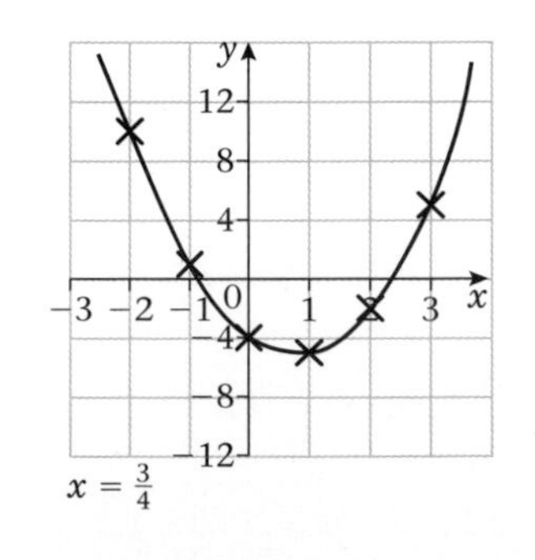

2 a $y = -1$ or -2 **b** $x = \frac{2}{3}$ or -5

c $x = -\frac{1}{5}$ or 3 **d** $\frac{5 \pm \sqrt{7}}{2}$

3 a $\frac{-5 \pm \sqrt{17}}{2}$, -0.44 or -4.56

b $2 \pm \sqrt{7}$, 4.65 or -0.65

c $\frac{-3 \pm \sqrt{29}}{10}$, 0.24 or -0.84

d $\frac{5 \pm \sqrt{73}}{6}$, 2.25 or -0.59

4 a

b

c

d

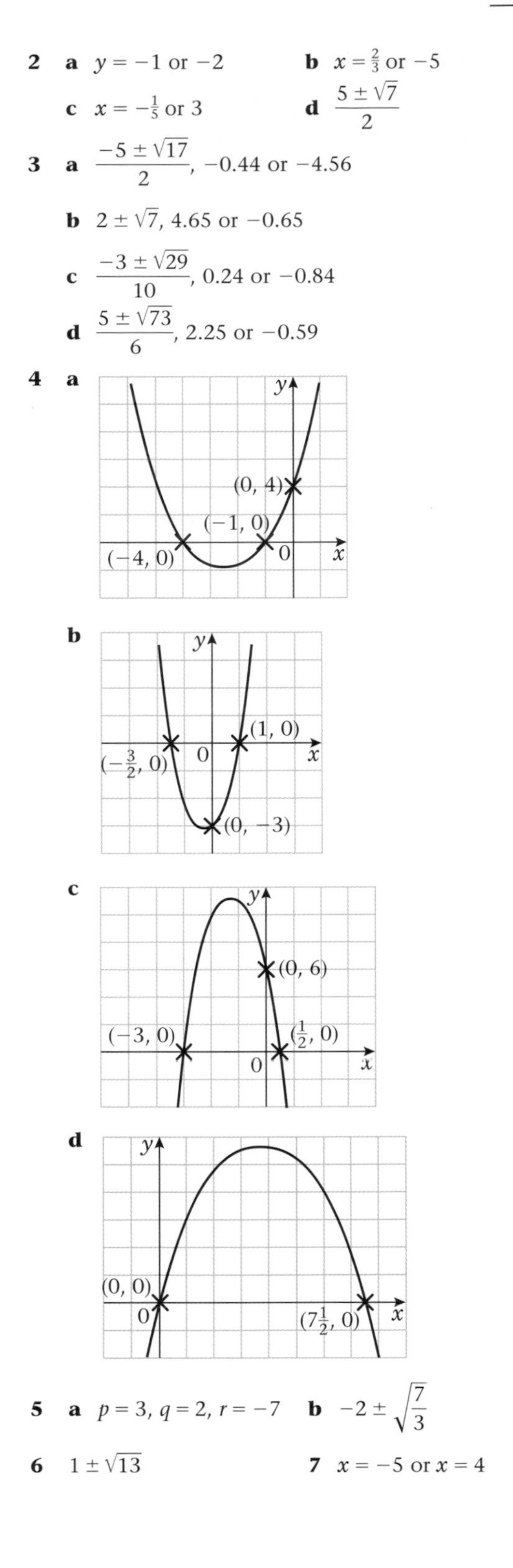

5 a $p = 3, q = 2, r = -7$ **b** $-2 \pm \sqrt{\frac{7}{3}}$

6 $1 \pm \sqrt{13}$ **7** $x = -5$ or $x = 4$

Exercise 3A

1 $x = 4, y = 2$
2 $x = 1, y = 3$
3 $x = 2, y = -2$
4 $x = 4\frac{1}{2}, y = -3$
5 $x = -\frac{2}{3}, y = 2$
6 $x = 3, y = 3$

Exercise 3B

1 $x = 5, y = 2$
2 $x = 5\frac{1}{2}, y = -6$
3 $x = 1, y = -4$
4 $x = 1\frac{3}{4}, y = \frac{1}{4}$

Exercise 3C

1 **a** $x = 5, y = 6$ or $x = 6, y = 5$
b $x = 0, y = 1$ or $x = \frac{4}{5}, y = -\frac{3}{5}$
c $x = -1, y = -3$ or $x = 1, y = 3$
d $x = 4\frac{1}{2}, y = 4\frac{1}{2}$ or $x = 6, y = 3$
e $a = 1, b = 5$ or $a = 3, b = -1$
f $u = 1\frac{1}{2}, v = 4$ or $u = 2, v = 3$
2 $(-11, -15)$ and $(3, -1)$
3 $(-1\frac{1}{6}, -4\frac{1}{2})$ and $(2, 5)$
4 **a** $x = -1\frac{1}{2}, y = 5\frac{3}{4}$ or $x = 3, y = -1$
b $x = 3, y = \frac{1}{2}$ or $x = 6\frac{1}{3}, y = -2\frac{5}{6}$
5 **a** $x = 3 + \sqrt{13}, y = -3 + \sqrt{13}$ or $x = 3 - \sqrt{13}$, $y = -3 - \sqrt{13}$
b $x = 2 - 3\sqrt{5}, y = 3 + 2\sqrt{5}$ or $x = 2 + 3\sqrt{5}$, $y = 3 - 2\sqrt{5}$

Exercise 3D

1 **a** $x < 4$ **b** $x \geqslant 7$ **c** $x > 2\frac{1}{2}$
d $x \leqslant -3$ **e** $x < 11$ **f** $x < 2\frac{3}{5}$
g $x > -12$ **h** $x < 1$ **i** $x \leqslant 8$
j $x > 1\frac{1}{7}$
2 **a** $x \geqslant 3$ **b** $x < 1$ **c** $x \leqslant -3\frac{1}{4}$
d $x < 18$ **e** $x > 3$ **f** $x \geqslant 4\frac{2}{5}$
g $x < 4$ **h** $x > -7$ **i** $x \leqslant -\frac{1}{2}$
j $x \geqslant \frac{3}{4}$
3 **a** $x > 2\frac{1}{2}$ **b** $2 < x < 4$ **c** $2\frac{1}{2} < x < 3$
d No values **e** $x = 4$

Exercise 3E

1 **a** $3 < x < 8$ **b** $-4 < x < 3$
c $x < -2, x > 5$ **d** $x \leqslant -4, x \geqslant -3$
e $-\frac{1}{2} < x < 7$ **f** $x < -2, x > 2\frac{1}{2}$
g $\frac{1}{2} \leqslant x \leqslant 1\frac{1}{2}$ **h** $x < \frac{1}{3}, x > 2$
i $-3 < x < 3$ **j** $x < -2\frac{1}{2}, x > \frac{2}{3}$
k $x < 0, x > 5$ **l** $-1\frac{1}{2} \leqslant x \leqslant 0$
2 **a** $-5 < x < 2$ **b** $x < -1, x > 1$
c $\frac{1}{2} < x < 1$ **d** $-3 < x < \frac{1}{4}$
3 **a** $2 < x < 4$ **b** $x > 3$
c $-\frac{1}{4} < x < 0$ **d** No values
e $-5 < x < -3, x > 4$ **f** $-1 < x < 1, 2 < x < 3$
4 **a** $-2 < k < 6$ **b** $p \leqslant -8$ or $p \geqslant 0$

Mixed exercise 3F

1 $x = -4, y = 3\frac{1}{2}$
2 $(3, 1)$ and $(-2\frac{1}{5}, -1\frac{3}{5})$
3 **b** $x = 4, y = 3$ and $x = -2\frac{2}{3}, y = -\frac{1}{3}$
4 $x = -1\frac{1}{2}, y = 2\frac{1}{4}$ and $x = 4, y = -\frac{1}{2}$
5 **a** $x > 10\frac{1}{2}$ **b** $x < -2, x > 7$
6 $3 < x < 4$
7 **a** $x = -5, x = 4$ **b** $x < -5, x > 4$
8 **a** $x < 2\frac{1}{2}$
b $\frac{1}{2} < x < 5$
c $\frac{1}{2} < x < 2\frac{1}{2}$
9 $k \leqslant 3\frac{1}{5}$
10 $x < 0, x > 1$
11 **a** $1 \pm \sqrt{13}$ **b** $x < 1 - \sqrt{13}, x > 1 + \sqrt{13}$
12 **a** $x < -4, x > 9$ **b** $y < -3, y > 3$
13 **a** $2x + 2(x - 5) > 32$ **b** $x(x - 5) < 104$
c $10\frac{1}{2} < x < 13$

Exercise 4A

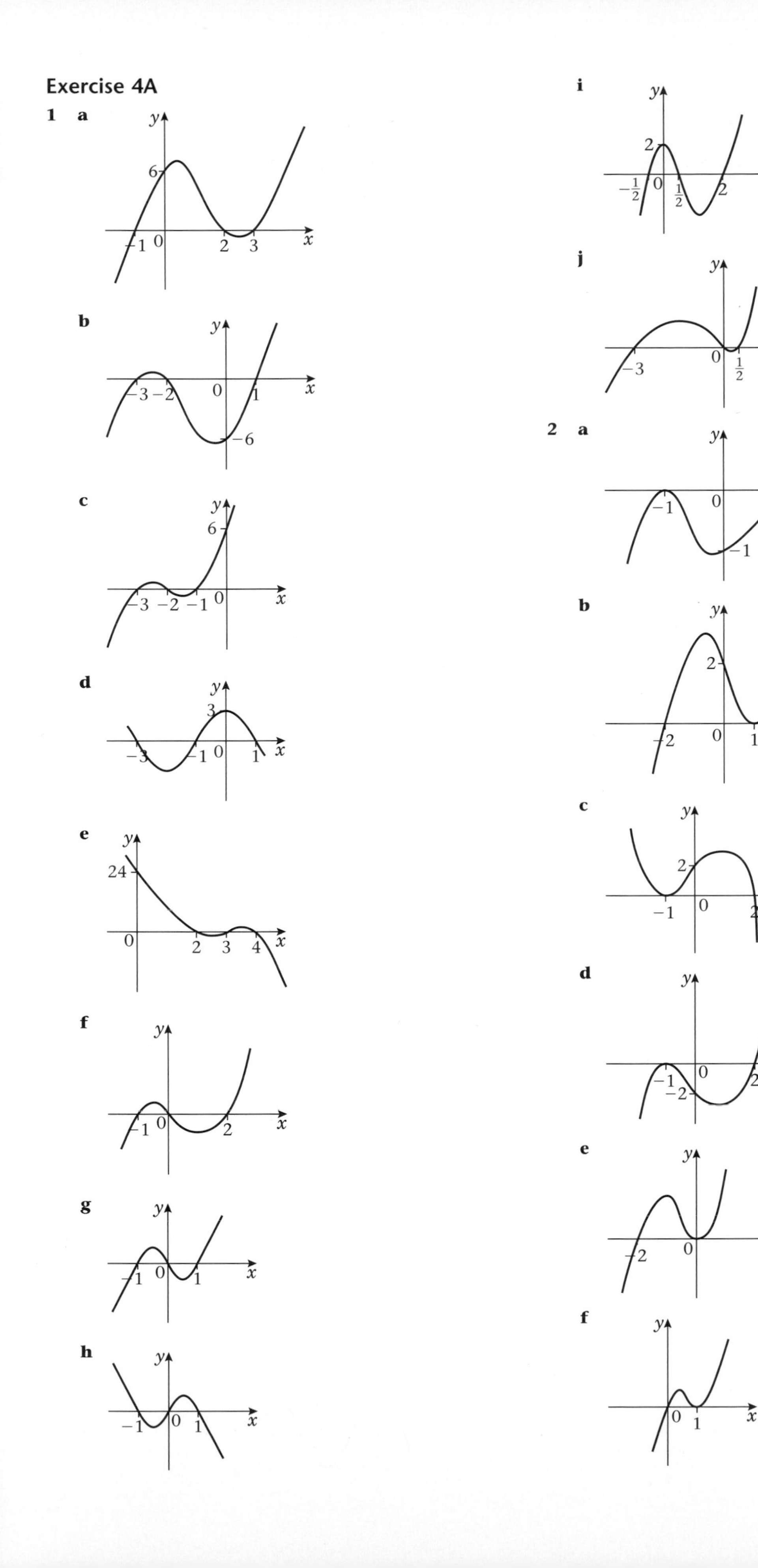

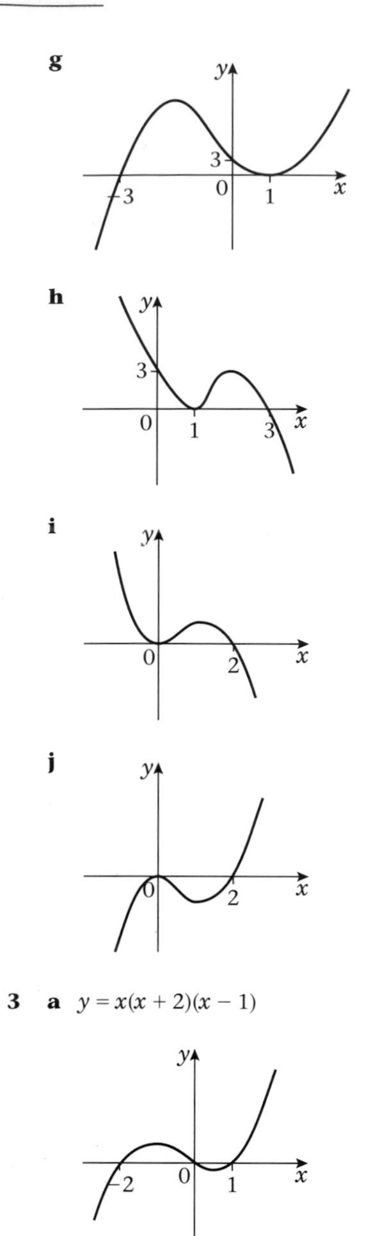
g
y
3
0
−3
1
x
h
y
3
0
1
3
x
i
y
0
2
x
j
y
0
2
x
3 a $y = x(x + 2)(x - 1)$
y
−2
0
1
x

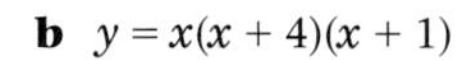
b $y = x(x + 4)(x + 1)$

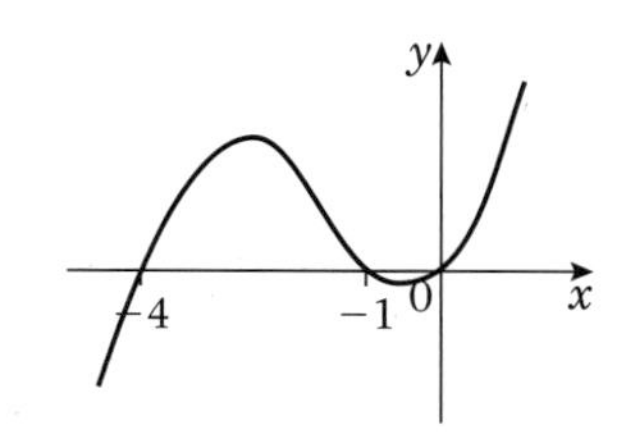
y
−4
−1
0
x

c $y = x(x + 1)^2$

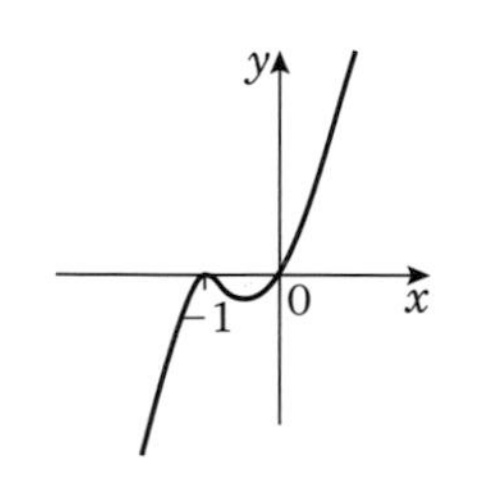
y
−1
0
x

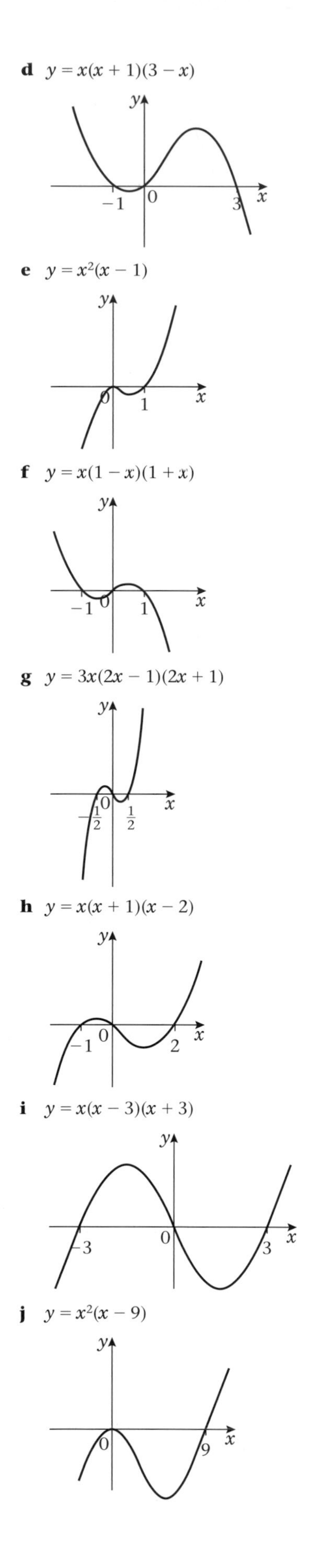
d $y = x(x + 1)(3 - x)$
y
−1
0
3
x
e $y = x^2(x - 1)$
y
0
1
x
f $y = x(1 - x)(1 + x)$
y
−1
0
1
x
g $y = 3x(2x - 1)(2x + 1)$
y
$-\frac{1}{2}$
0
$\frac{1}{2}$
x
h $y = x(x + 1)(x - 2)$
y
−1
0
2
x
i $y = x(x - 3)(x + 3)$
y
−3
0
3
x
j $y = x^2(x - 9)$
y
0
9
x

Exercise 4B

1 a, **b**, **c**, **d**, **e**; **2 a**

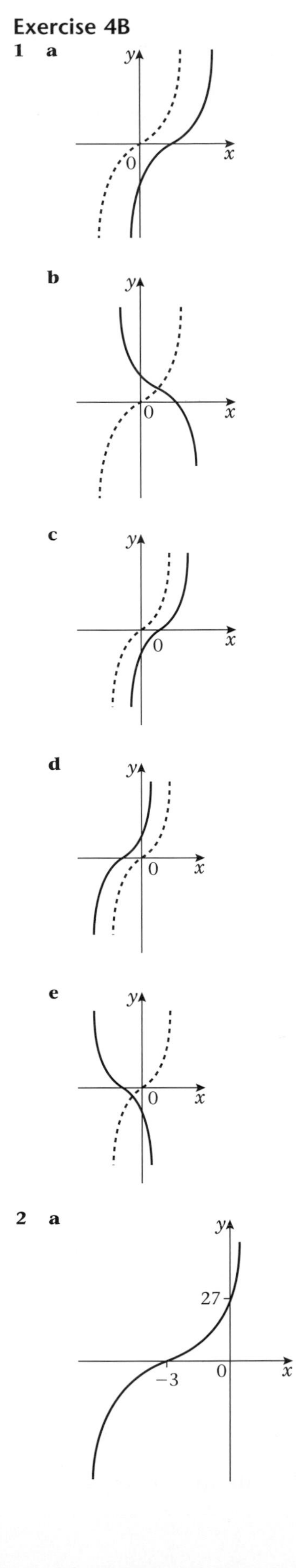

b, **c**, **d**, **e**

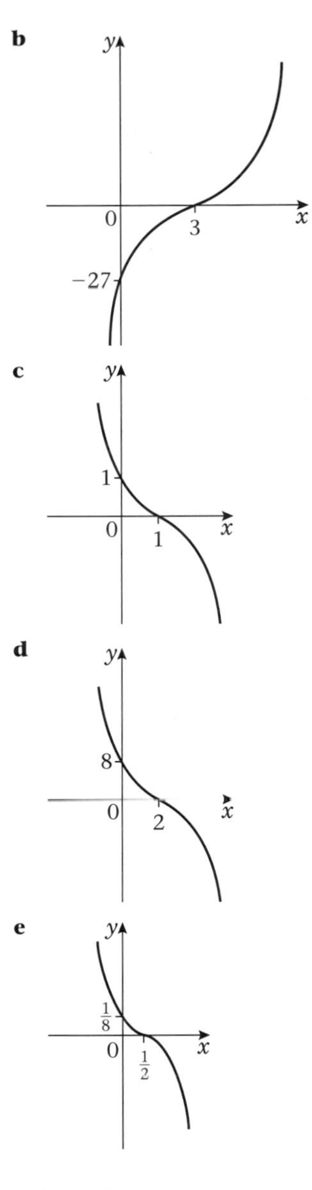

Exercise 4C

1

2

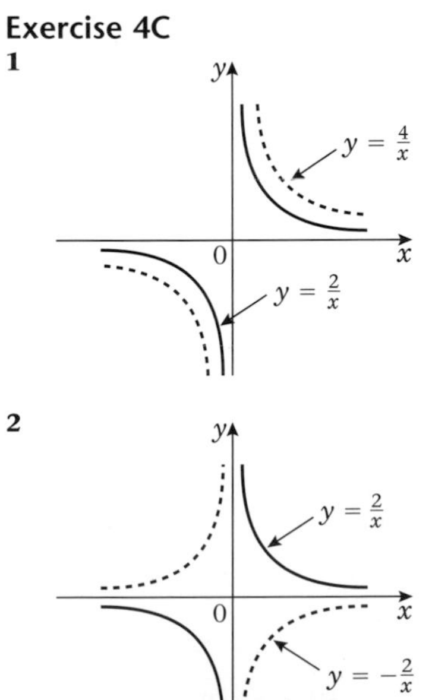

3

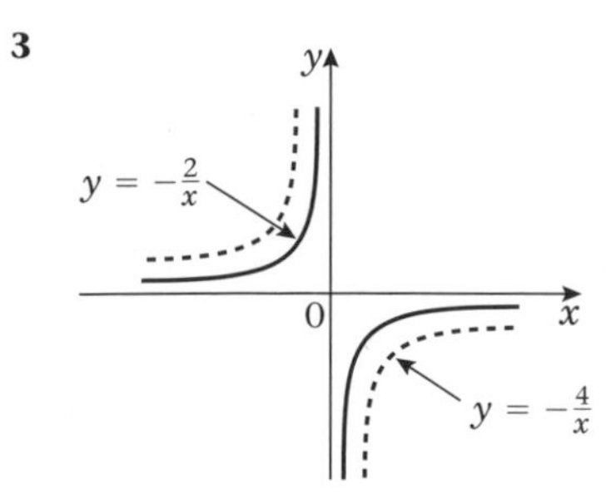

4

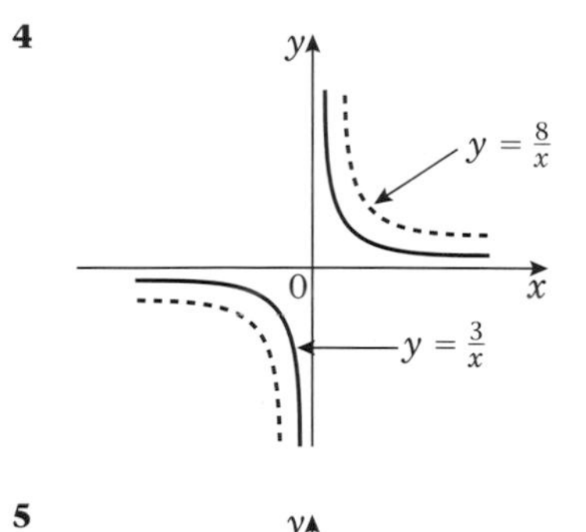

5

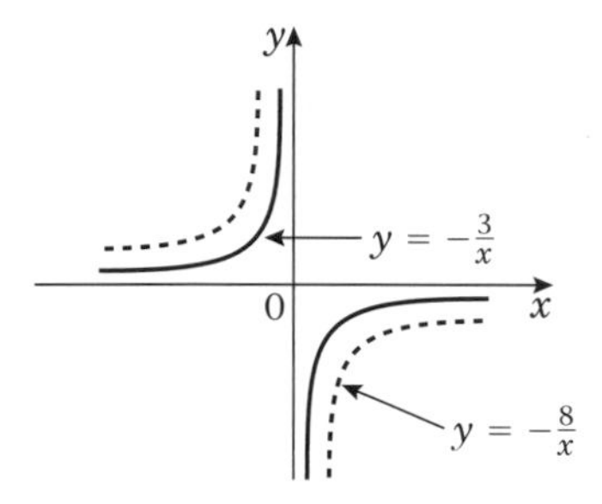

Exercise 4D

1 a i

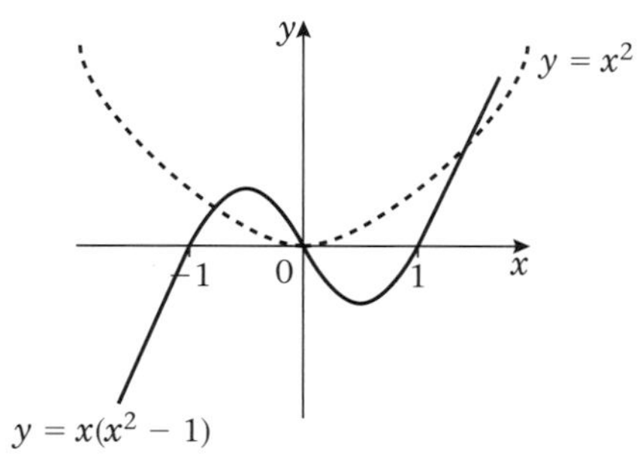

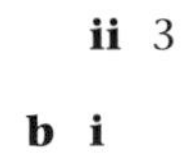

ii 3 **iii** $x^2 = x(x^2 - 1)$

b i

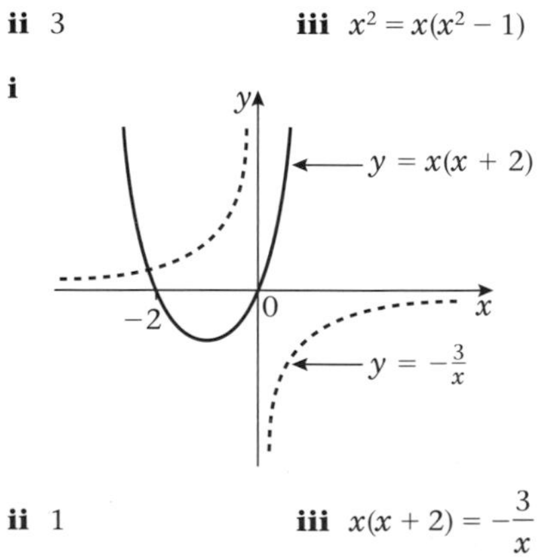

ii 1 **iii** $x(x + 2) = -\frac{3}{x}$

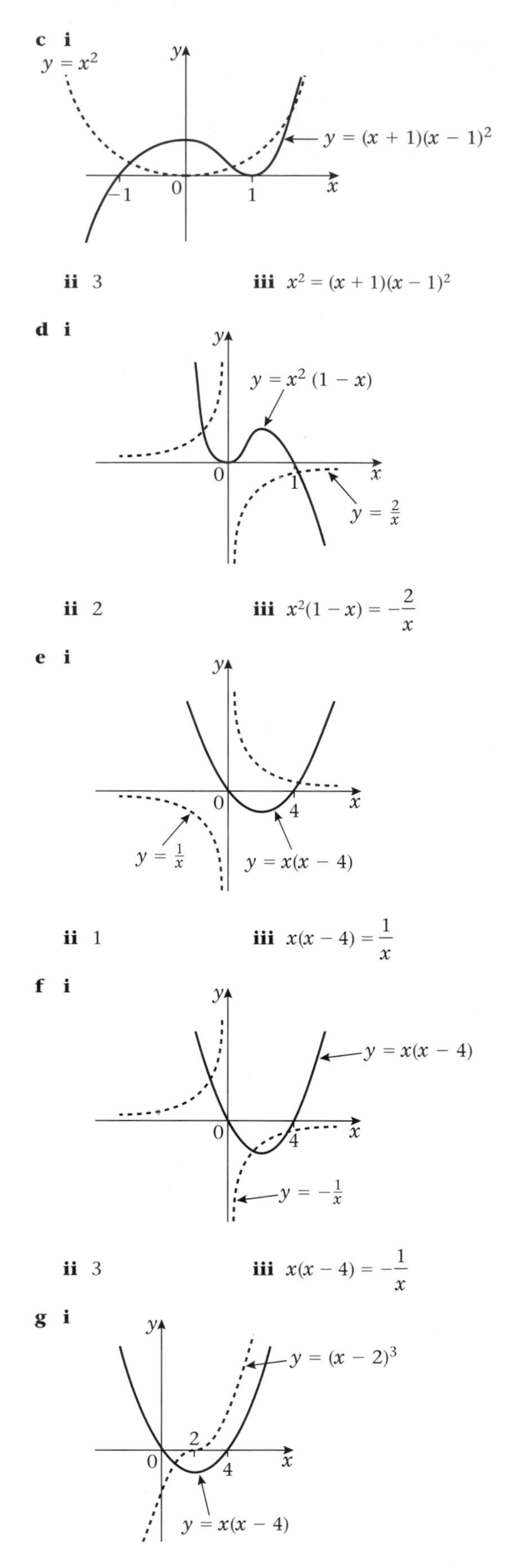

c i

ii 3 **iii** $x^2 = (x + 1)(x - 1)^2$

d i

ii 2 **iii** $x^2(1 - x) = -\frac{2}{x}$

e i

ii 1 **iii** $x(x - 4) = \frac{1}{x}$

f i

ii 3 **iii** $x(x - 4) = -\frac{1}{x}$

g i

ii 1 **iii** $x(x - 4) = (x - 2)^3$

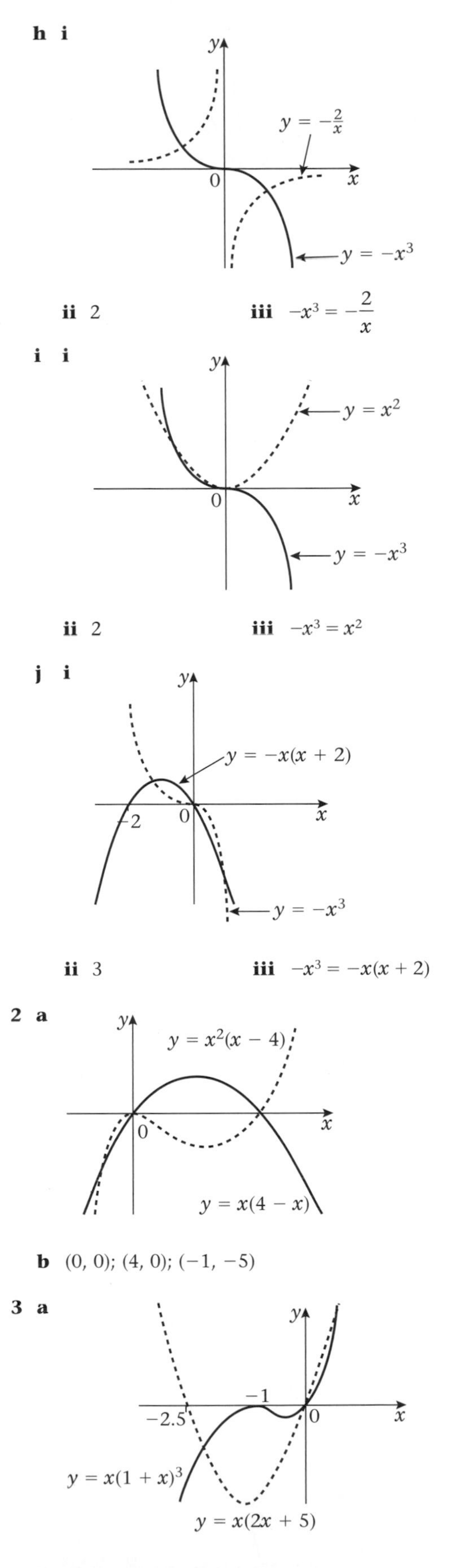

h i

ii 2 **iii** $-x^3 = -\frac{2}{x}$

i i

ii 2 **iii** $-x^3 = x^2$

j i

ii 3 **iii** $-x^3 = -x(x + 2)$

2 a

b (0, 0); (4, 0); (−1, −5)

3 a

b (0, 0); (2, 18); (−2, −2)

4 a

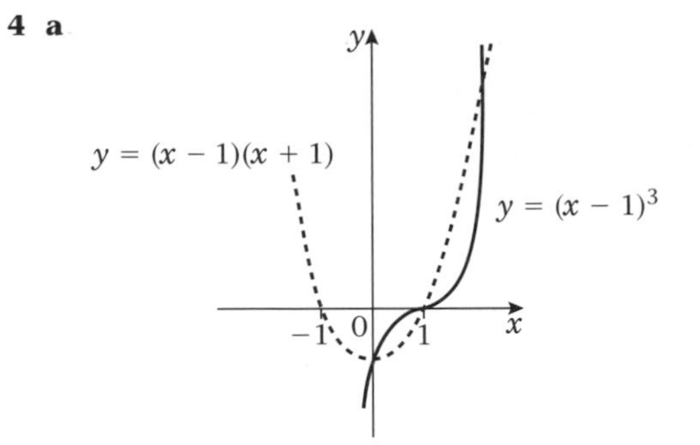

b (0, −1); (1, 0); (3, 8)

5 a

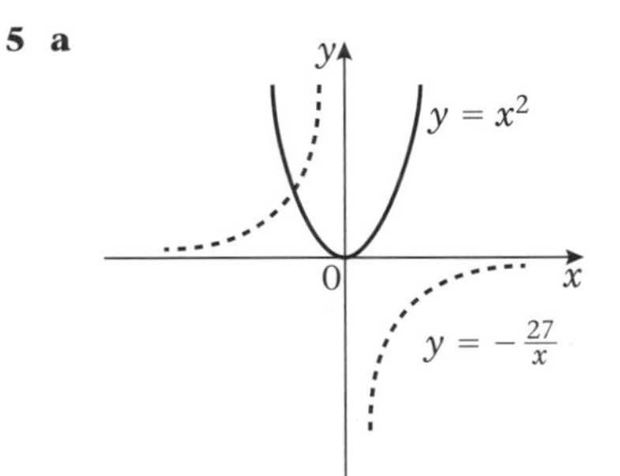

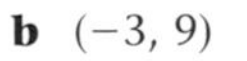

b (−3, 9)

6 a

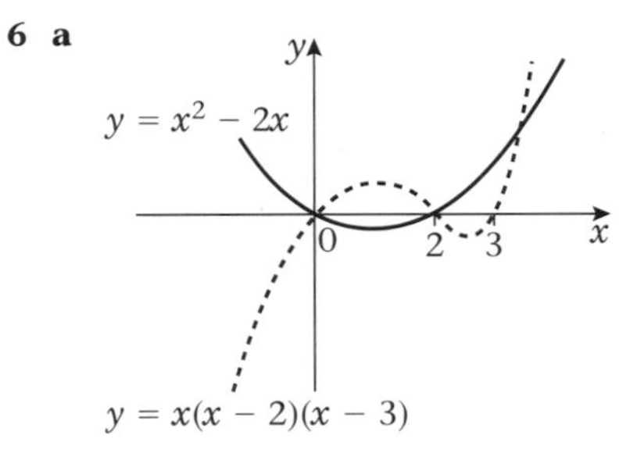

b (0, 0); (2, 0); (4, 8)

7 a

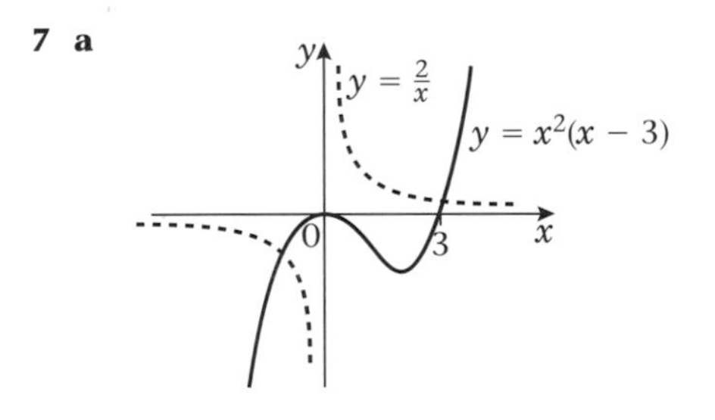

b Only 2 intersections

8 a

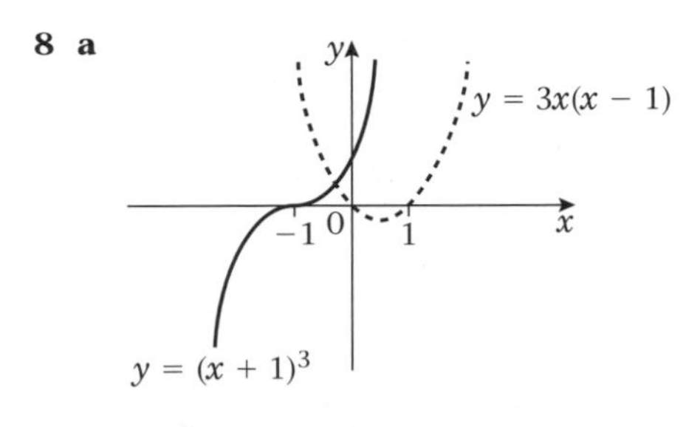

b Only 1 intersection

9 a

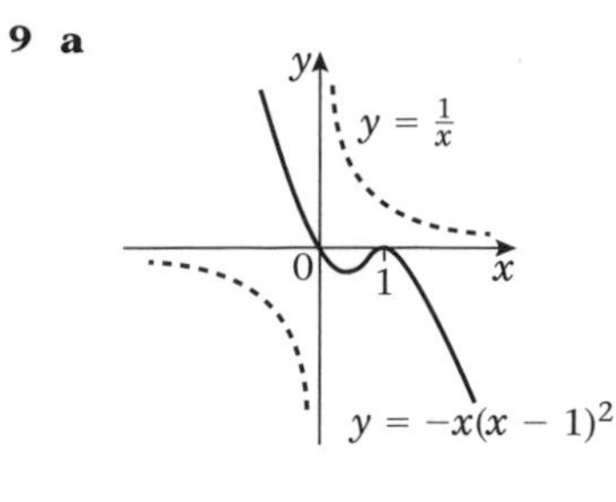

b Graphs do not intersect.

10 a

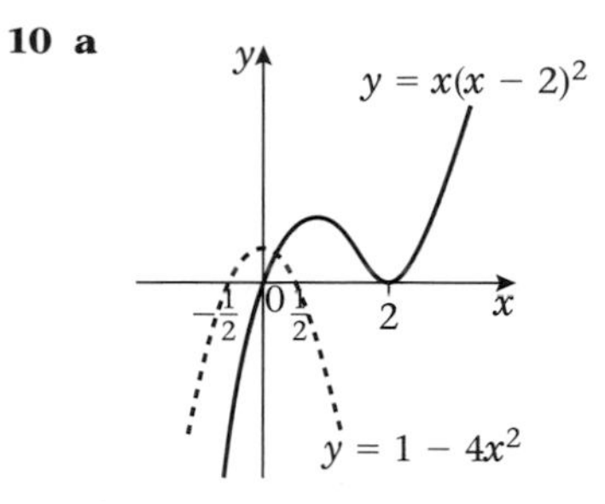

b 1, since graphs only cross once

11 a

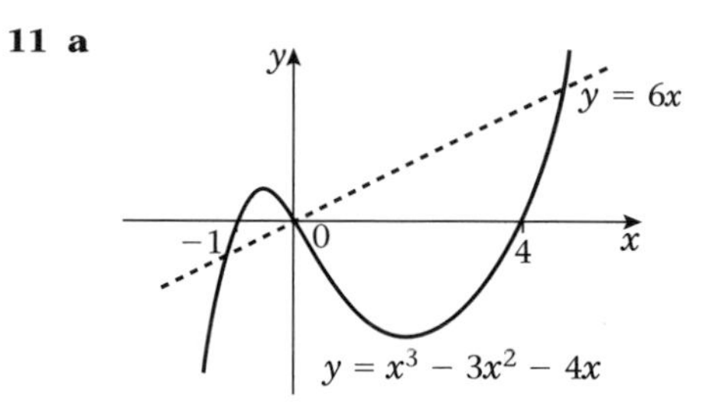

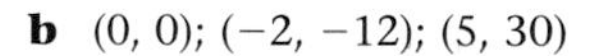

b (0, 0); (−2, −12); (5, 30)

12 a

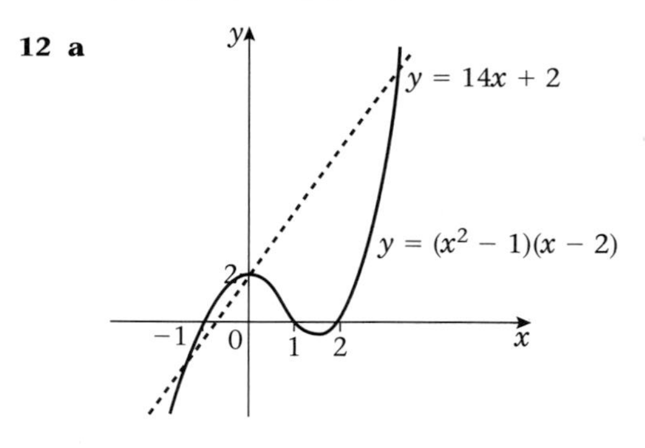

b (0, 2); (−3, −40); (5, 72)

13 a

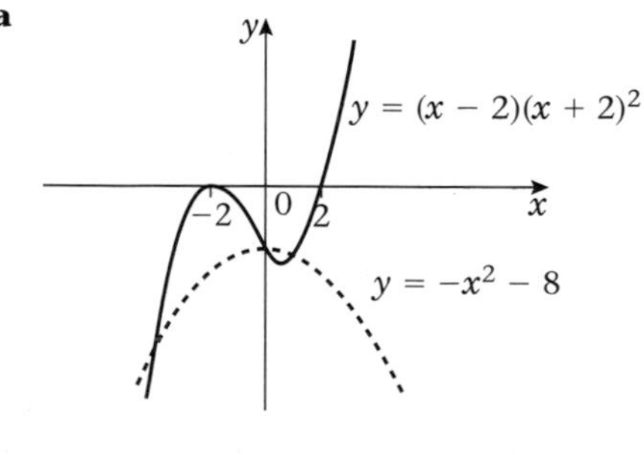

b (0, −8); (1, −9); (−4, −24)

Exercise 4E

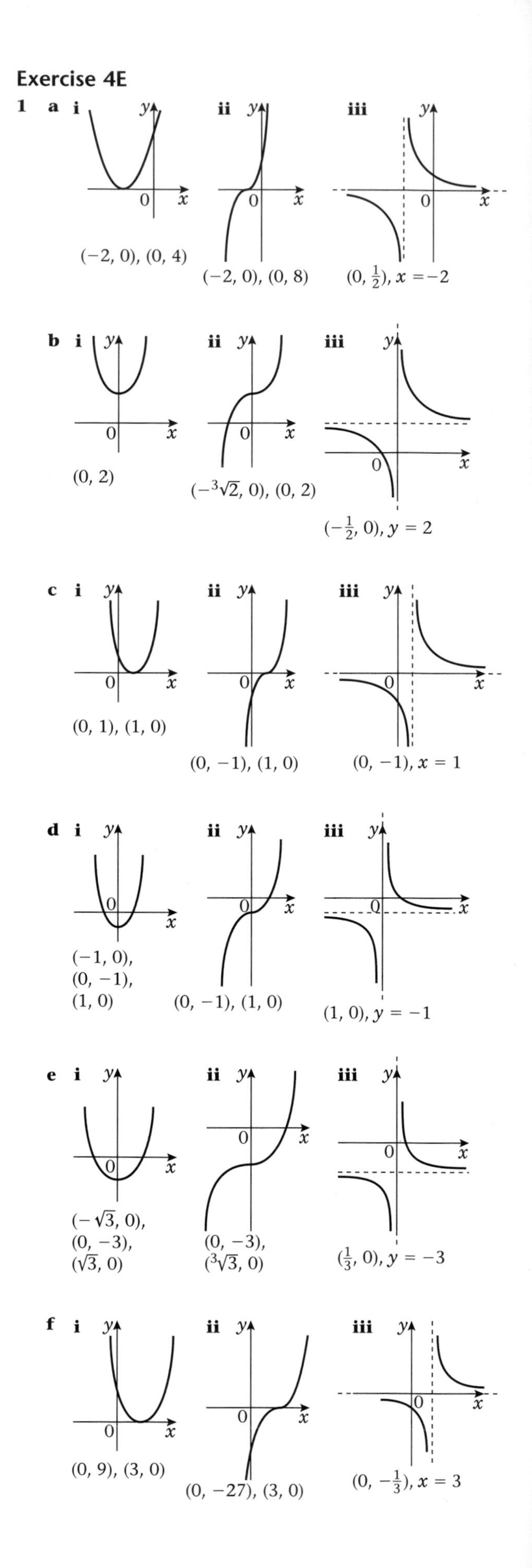

2 **a**

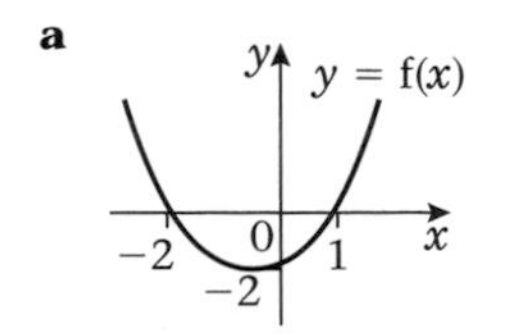

b **i**

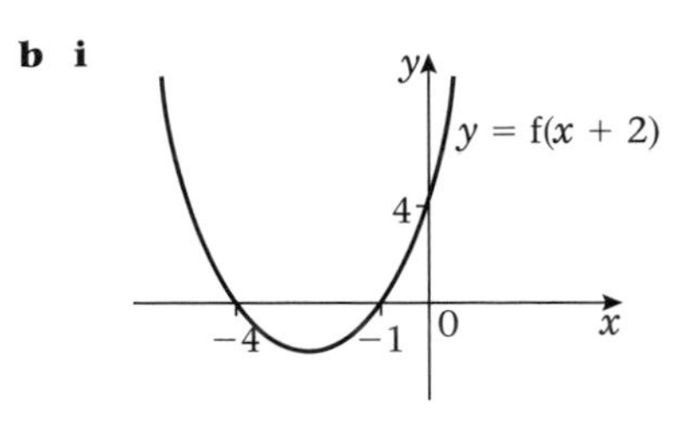

ii

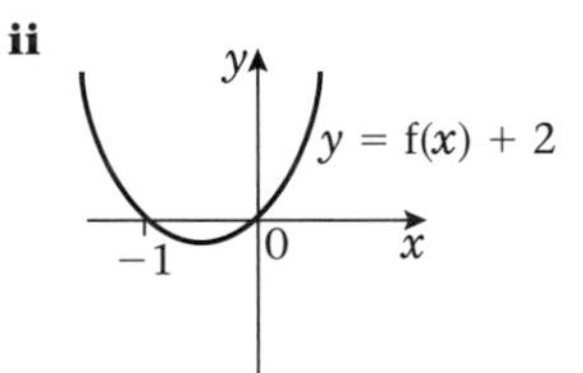

c $f(x + 2) = (x + 1)(x + 4)$; $(0, 4)$
$f(x) + 2 = (x - 1)(x + 2) + 2$; $(0, 0)$

3 **a**

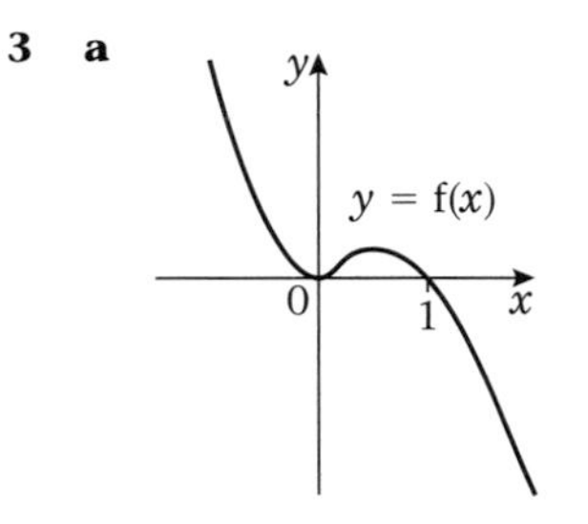

b

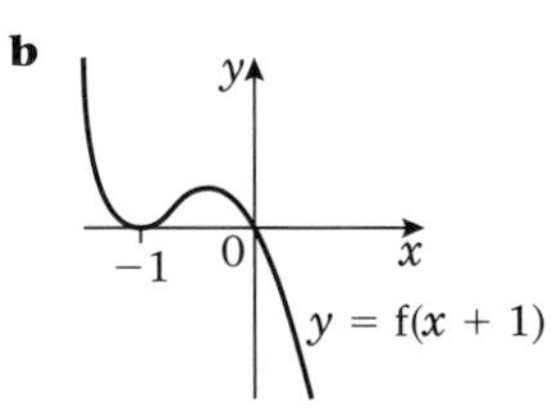

c $f(x + 1) = -x(x + 1)^2$; $(0, 0)$

4 **a**

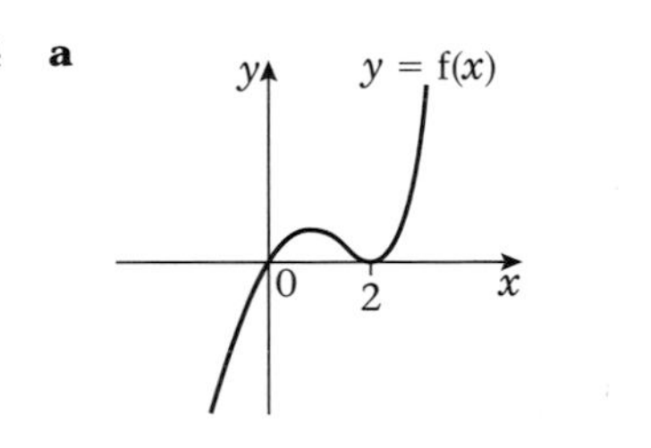

b

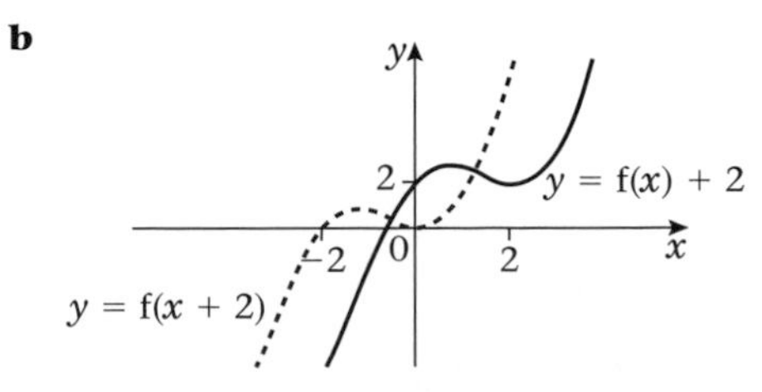

c $f(x + 2) = (x + 2)x^2$; $(0, 0)$; $(-2, 0)$

5 **a**

b

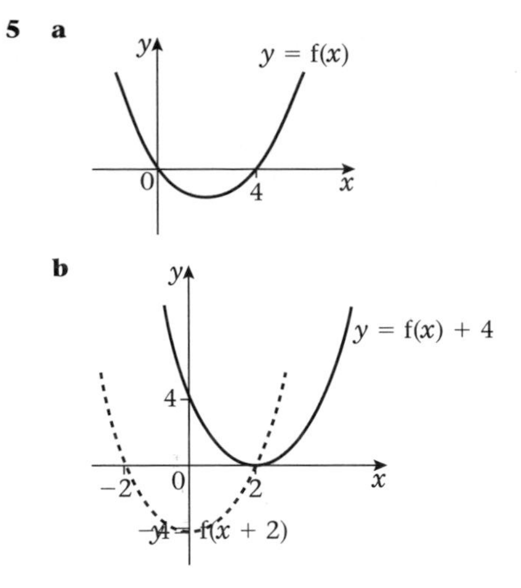

c $f(x + 2) = (x + 2)(x - 2)$; $(2, 0)$; $(-2, 0)$
$f(x) + 4 = (x - 2)^2$; $(2, 0)$

Exercise 4F

1 **a** **i** **ii** **iii**
b **i** **ii** **iii**
c **i** **ii** **iii**
d **i** **ii** **iii**
e **i** **ii** **iii**

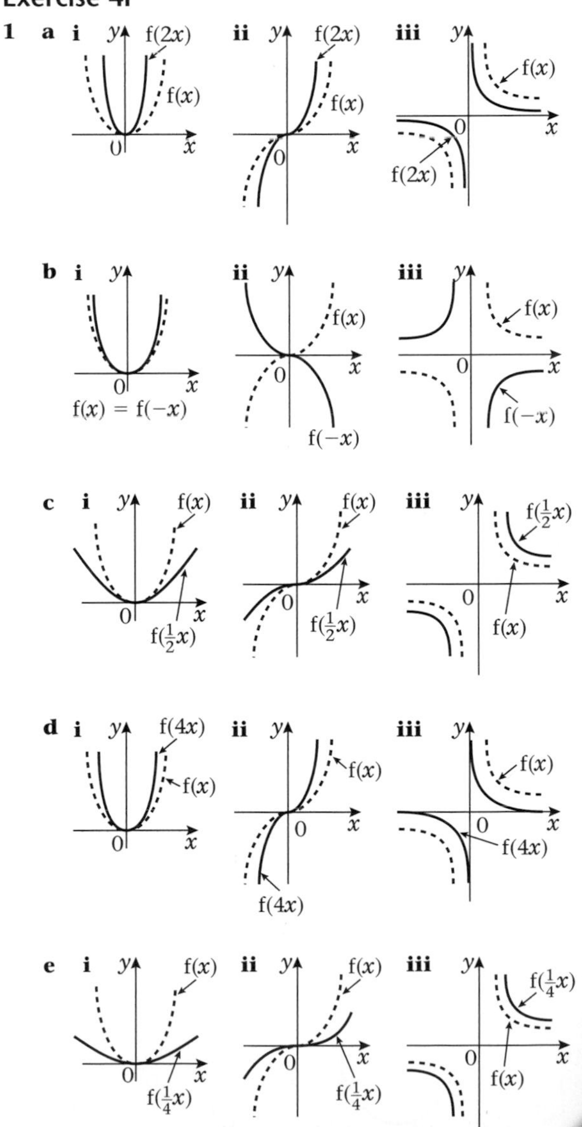

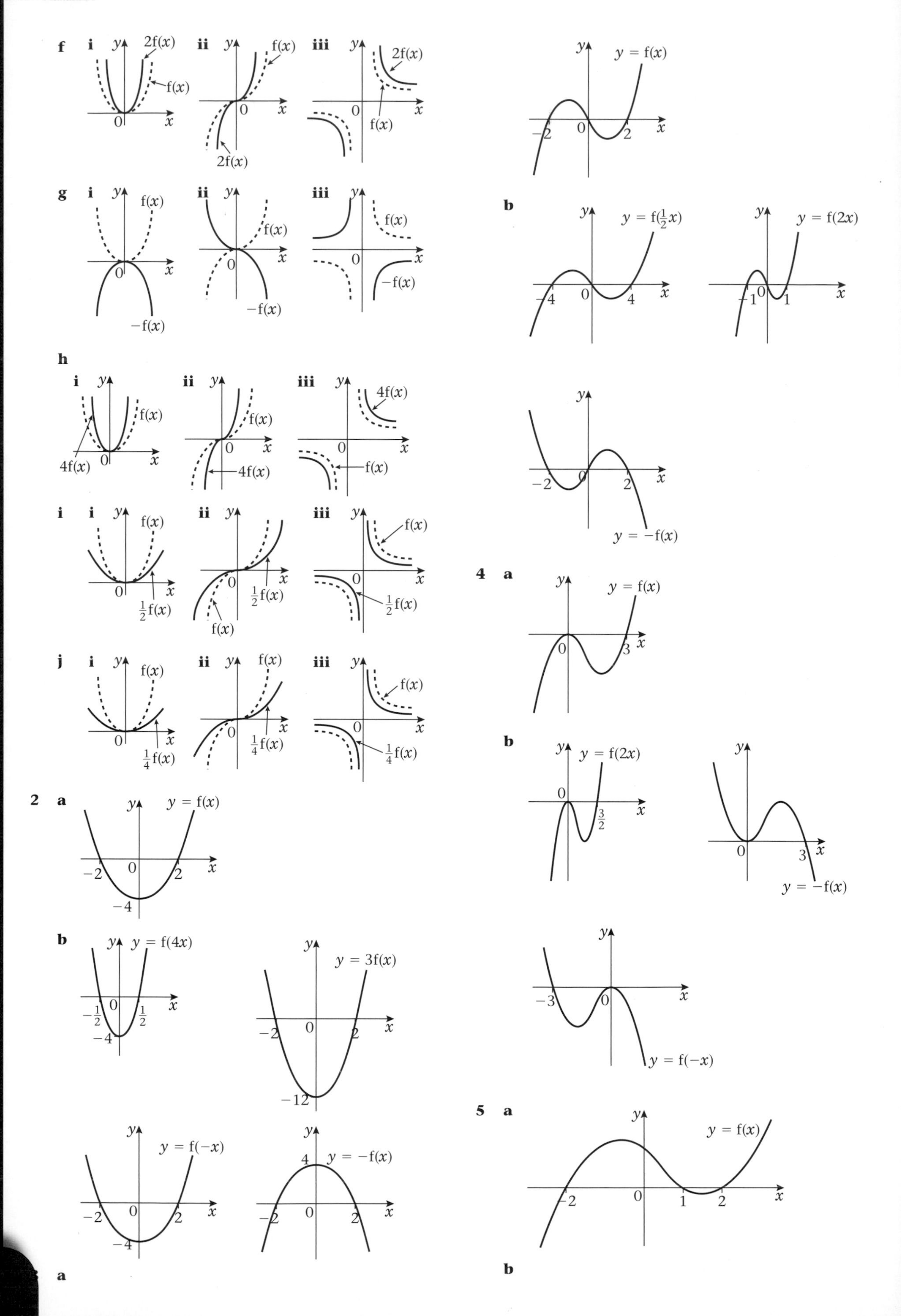
f
i
y
2f(x)
f(x)
0
x
ii
f(x)
2f(x)
iii
2f(x)
g
i
f(x)
−f(x)
ii
f(x)
−f(x)
iii
f(x)
−f(x)
h
i
f(x)
4f(x)
ii
f(x)
−4f(x)
iii
4f(x)
−f(x)
i
i
f(x)
$\frac{1}{2}$f(x)
ii
$\frac{1}{2}$f(x)
f(x)
iii
f(x)
$\frac{1}{2}$f(x)
j
i
f(x)
$\frac{1}{4}$f(x)
ii
f(x)
$\frac{1}{4}$f(x)
iii
f(x)
$\frac{1}{4}$f(x)
2 a
y = f(x)
−2
2
−4
b
y = f(4x)
$-\frac{1}{2}$
$\frac{1}{2}$
−4
y = 3f(x)
−2
2
−12
y = f(−x)
−2
2
−4
4
y = −f(x)
−2
2
a
y = f(x)
−2
2
b
y = f($\frac{1}{2}$x)
−4
4
y = f(2x)
−1
1
y = −f(x)
−2
2
4 a
y = f(x)
3
b
y = f(2x)
$\frac{3}{2}$
y = −f(x)
3
−3
y = f(−x)
5 a
y = f(x)
−2
1
2
b

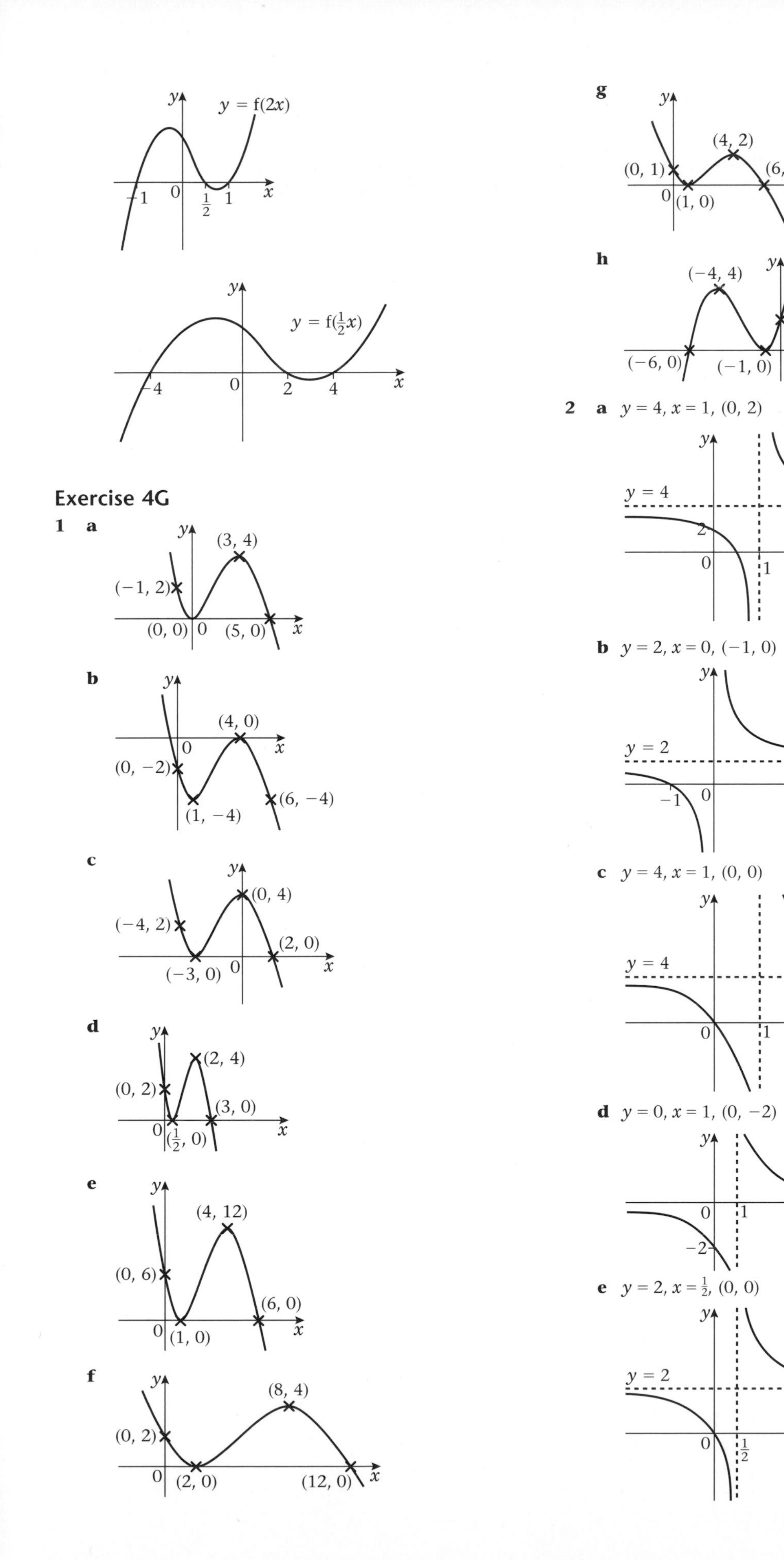

Exercise 4G

1 **a**

b

c

d

e

f

g

h

2 **a** $y = 4$, $x = 1$, $(0, 2)$

b $y = 2$, $x = 0$, $(-1, 0)$

c $y = 4$, $x = 1$, $(0, 0)$

d $y = 0$, $x = 1$, $(0, -2)$

e $y = 2$, $x = \frac{1}{2}$, $(0, 0)$

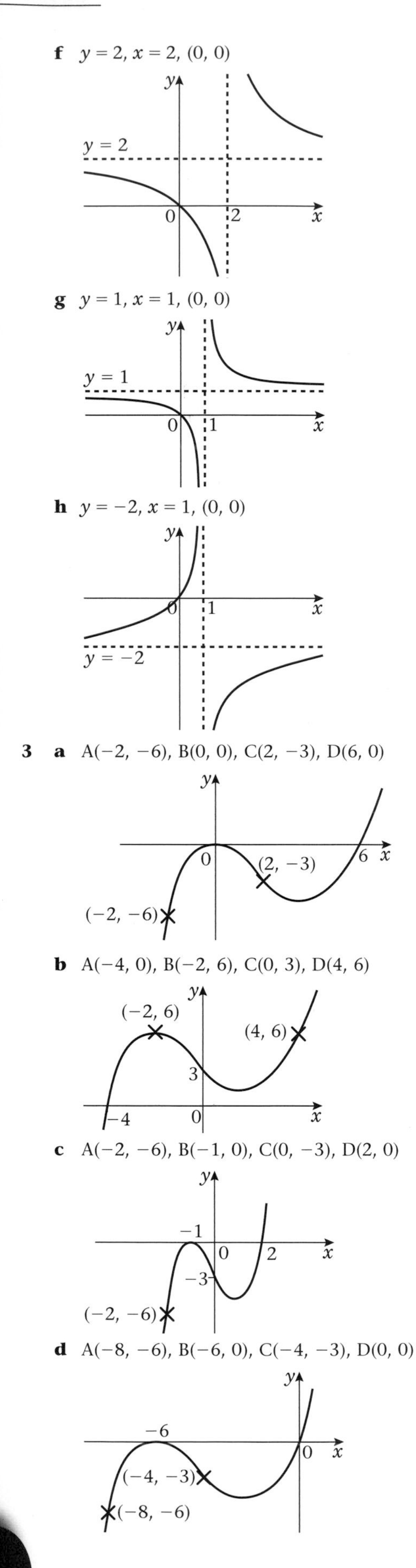
f $y = 2, x = 2, (0, 0)$
g $y = 1, x = 1, (0, 0)$
h $y = -2, x = 1, (0, 0)$
3 a A(−2, −6), B(0, 0), C(2, −3), D(6, 0)
b A(−4, 0), B(−2, 6), C(0, 3), D(4, 6)
c A(−2, −6), B(−1, 0), C(0, −3), D(2, 0)
d A(−8, −6), B(−6, 0), C(−4, −3), D(0, 0)

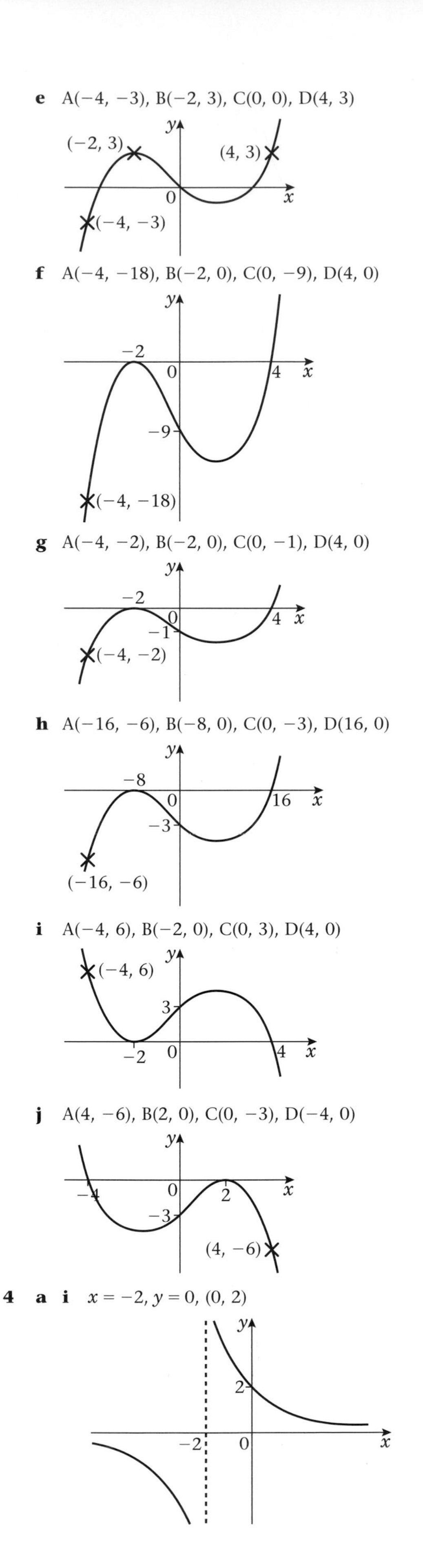
e A(−4, −3), B(−2, 3), C(0, 0), D(4, 3)
f A(−4, −18), B(−2, 0), C(0, −9), D(4, 0)
g A(−4, −2), B(−2, 0), C(0, −1), D(4, 0)
h A(−16, −6), B(−8, 0), C(0, −3), D(16, 0)
i A(−4, 6), B(−2, 0), C(0, 3), D(4, 0)
j A(4, −6), B(2, 0), C(0, −3), D(−4, 0)
4 a i $x = -2, y = 0, (0, 2)$

ii $x = -1$, $y = 0$, $(0, 1)$

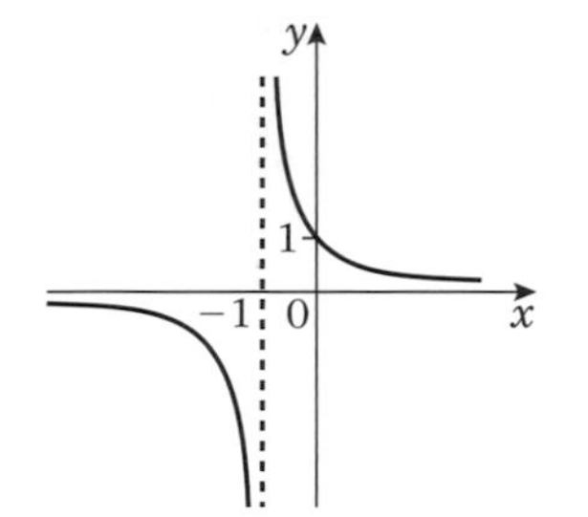

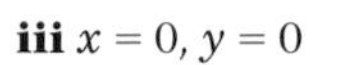

iii $x = 0$, $y = 0$

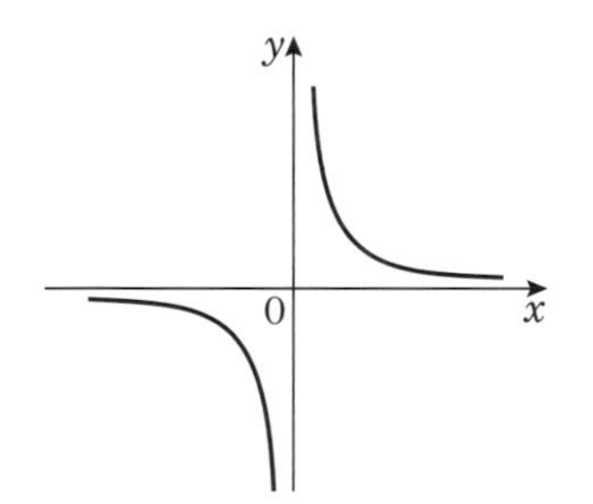

iv $x = -2$, $y = -1$, $(0, 0)$

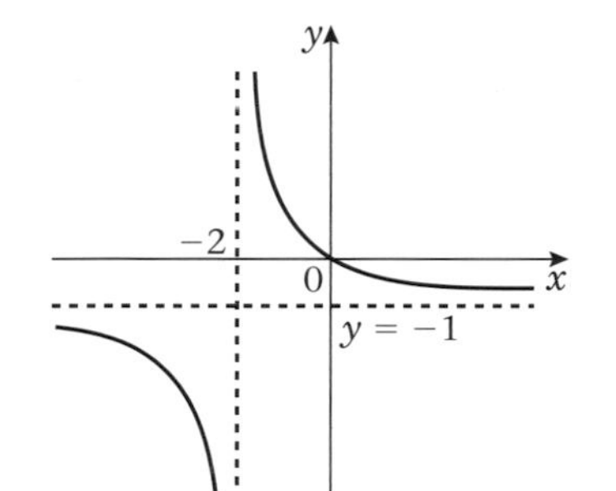

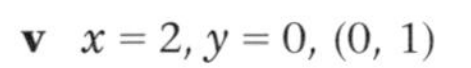

v $x = 2$, $y = 0$, $(0, 1)$

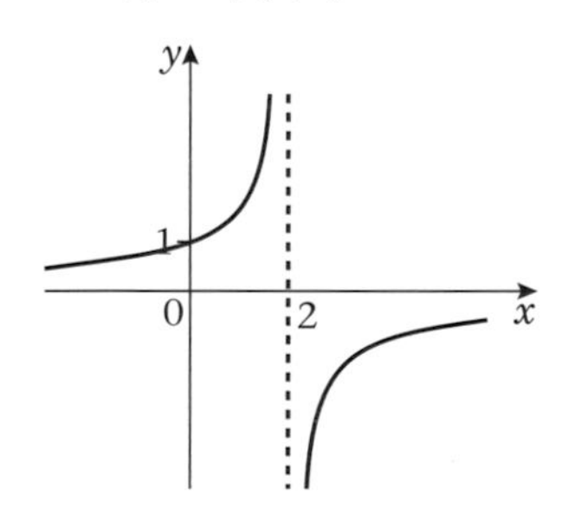

vi $x = -2$, $y = 0$, $(0, -1)$

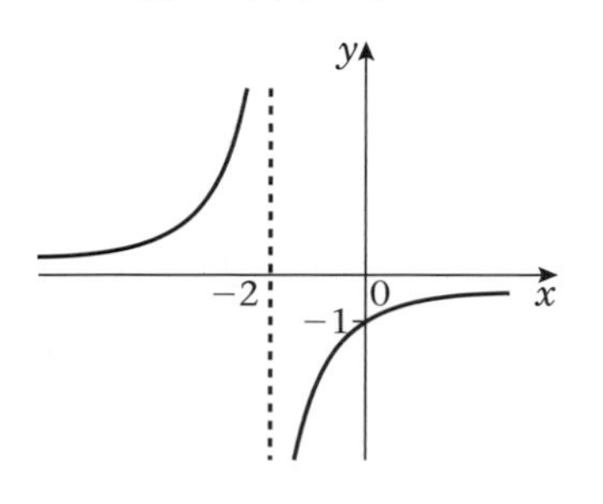

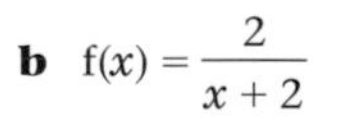

b $f(x) = \dfrac{2}{x + 2}$

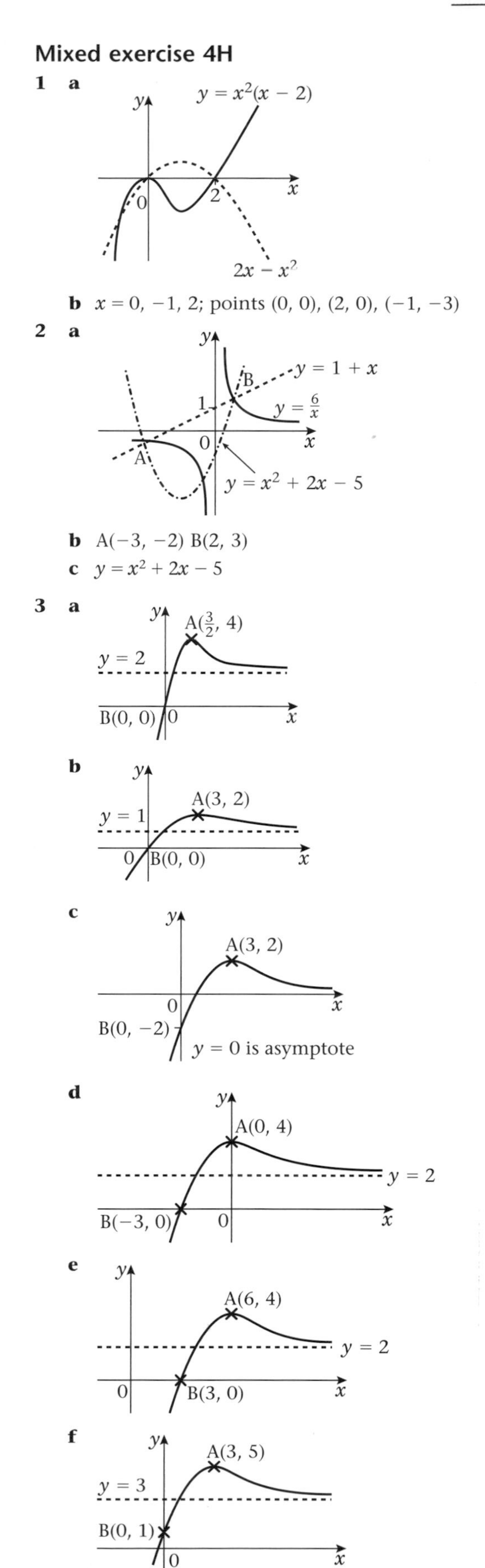

Mixed exercise 4H

1 a

b $x = 0, -1, 2$; points $(0, 0)$, $(2, 0)$, $(-1, -3)$

2 a

b A$(-3, -2)$ B$(2, 3)$

c $y = x^2 + 2x - 5$

3 a

b

c

d

e

f

4 a $x = -1$ at A, $x = 3$ at B

5 a

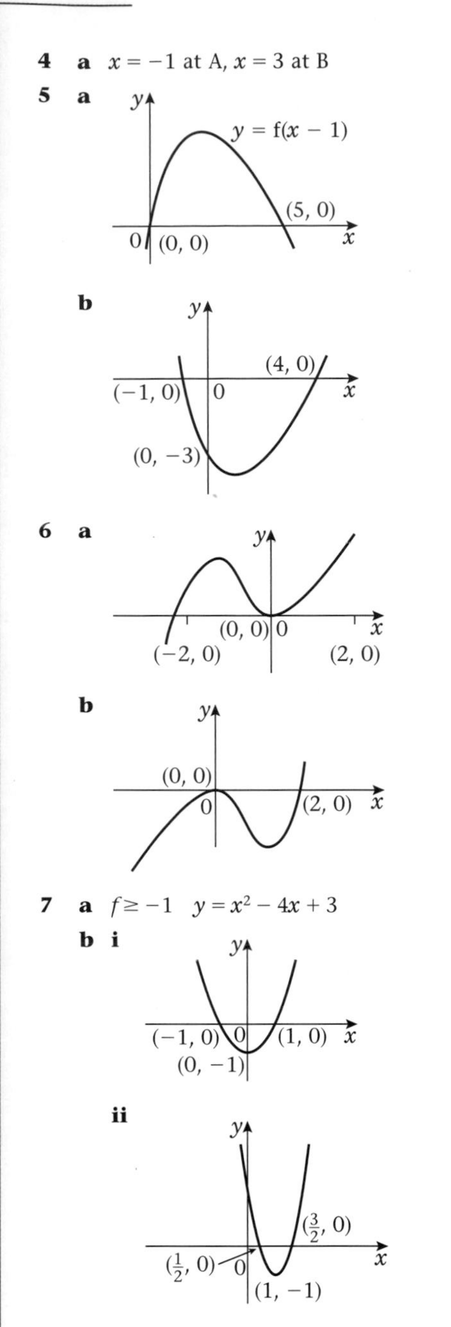

7 a $f \geq -1 \quad y = x^2 - 4x + 3$

Review Exercise 1

1 **a** $x(2x + 1)(x - 7)$
  b $(3x - 4)(3x + 4)$
  c $(x + 1)(x - 1)(x^2 + 8)$

2 **a** 9 **b** 27 **c** $\frac{1}{27}$

3 **a** 2 **b** $\frac{1}{4}$

4 **a** 625 **b** $\frac{4}{3}x^{\frac{2}{3}}$

5 **a** $4\sqrt{5}$ **b** $21 - 8\sqrt{5}$

6 **a** 13 **b** $8 - 2\sqrt{3}$

7 **a** $6\sqrt{3}$ **b** $7 - 4\sqrt{3}$

8 **a** $56\sqrt{7}$ **b** $10 - 13\sqrt{7}$ **c** $16 + 6\sqrt{7}$

9 **a** $x = -8$ or $x = 9$
  b $x = 0$ or $x = -\frac{7}{2}$
  c $x = -\frac{3}{2}$ or $x = \frac{3}{5}$

10 **a** $x = -2.17$ or -7.83
  b $x = 2.69$ or $x = -0.186$
  c $x = 2.82$ or $x = 0.177$

11 **a** $a = -4$, $b = -45$ **b** $x = 4 \pm 3\sqrt{5}$

12 **a** $(x - 3)^2 + 9$
  b P is (0, 18), Q is (3, 9)
  c $x = 3 + 4\sqrt{2}$

13 $k = 6$, $x = -1$ (same root)

14 **a** $a = 5$, $b = 11$
  b discriminant < 0 so no real roots
  c $k = 25$
  d

y
25
−5 0 x

15 **a** $a = 1$, $b = 2$
  b

y
3
0 x

  c discriminant = −8
  d $-2\sqrt{3}\ k < 2\sqrt{3}$

16 $y = 4$, $x = -2$ or $y = -2$, $x = 4$

17 **a** $x^2 + 4x - 8 = 0$
  b $x = -2 \pm 2\sqrt{3}$, $y = -6 \pm 2\sqrt{3}$

18 $x = 2$, $y = -1$ or $x = -\frac{1}{3}$, $y = -\frac{17}{3}$

19 **a** $x > \frac{1}{4}$
  b $x < \frac{1}{2}$ or $x > 3$
  c $\frac{1}{4} < x < \frac{1}{2}$ or $x > 3$

20 **a** $0 < x < 6$
  b $x < -4$ or $x > \frac{5}{3}$

21 **a** $x = \frac{7}{2}$, $y = -2$, $x = -3$, $y = 11$
  b $x < -3$ or $x > 3\frac{1}{2}$

22 **a** Different real roots, determinant > 0, so $k^2 - 4k - 12 > 0$
  b $k < -2$ or $k > 6$

23 $0 < k < \frac{8}{9}$

24 **a** $p^2 - 8p - 20 > 0$
  b $p < -2$ or $p > 10$
  c $x = \dfrac{(-3 \pm \sqrt{13})}{2}$

25 **a** $x(x - 2)(x + 2)$
  b
  c

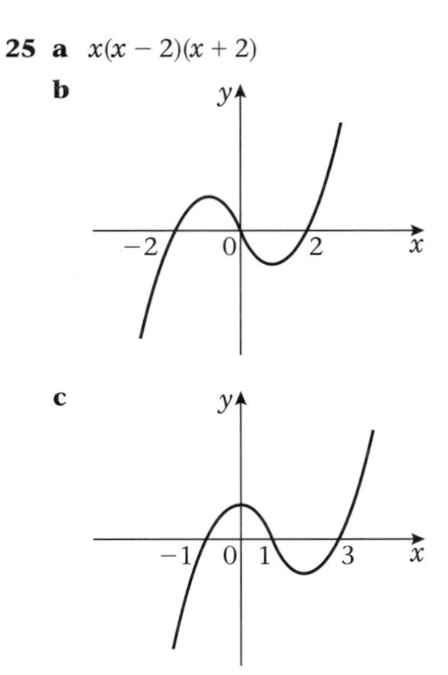

26 **a**

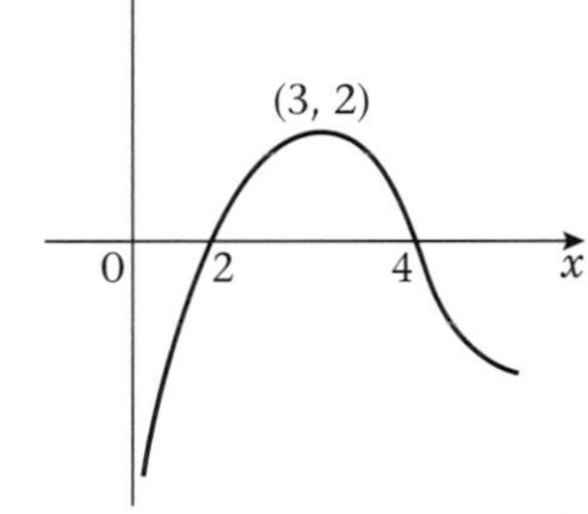

(2, 0) (4, 0) and (3, 2)

  b

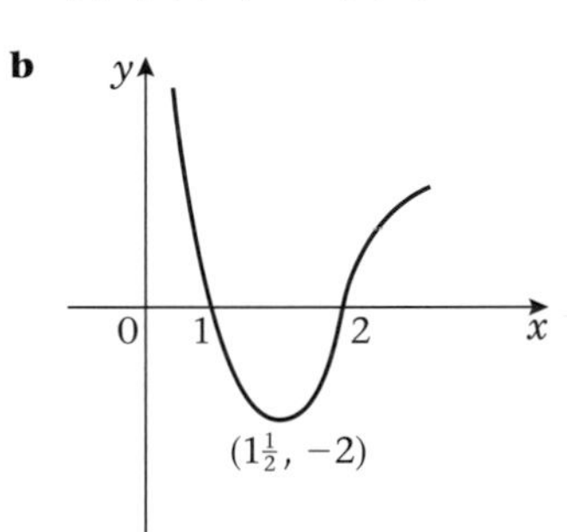

(1, 0) (2, 0) and $(1\frac{1}{2}, -2)$

27 **a**

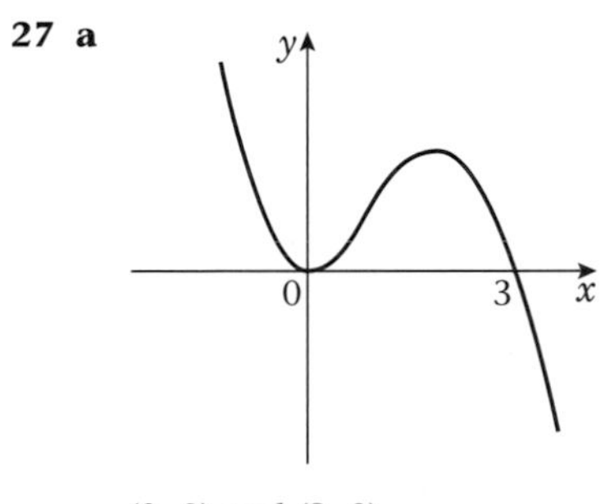

(0, 0) and (3, 0)

b

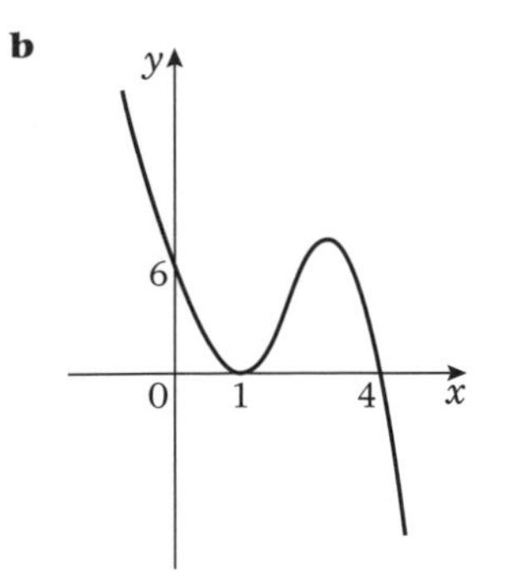

(1, 0) (4, 0) and (0, 6)

c

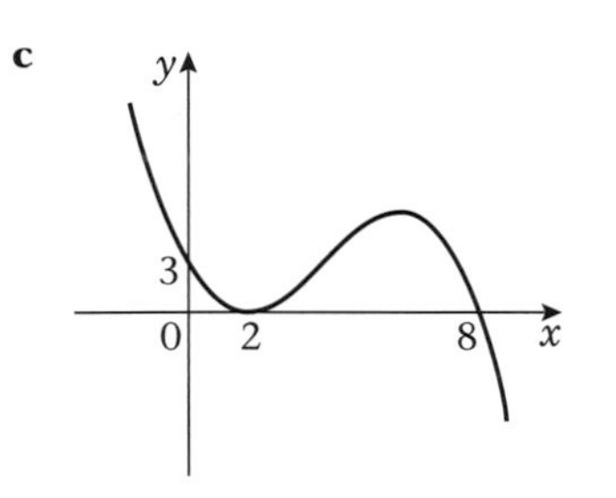

(2, 0) (8, 0) and (0, 3)

28 a

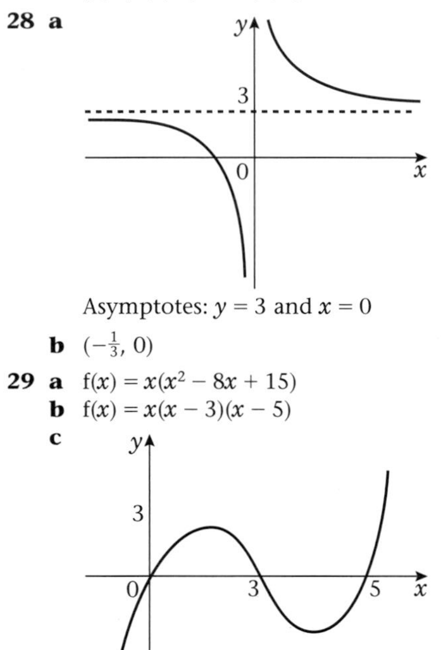

Asymptotes: $y = 3$ and $x = 0$

b $(-\frac{1}{3}, 0)$

29 a $f(x) = x(x^2 - 8x + 15)$

b $f(x) = x(x - 3)(x - 5)$

c

(0, 0), (3, 0) and (5, 0)

30 a

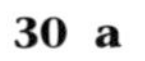

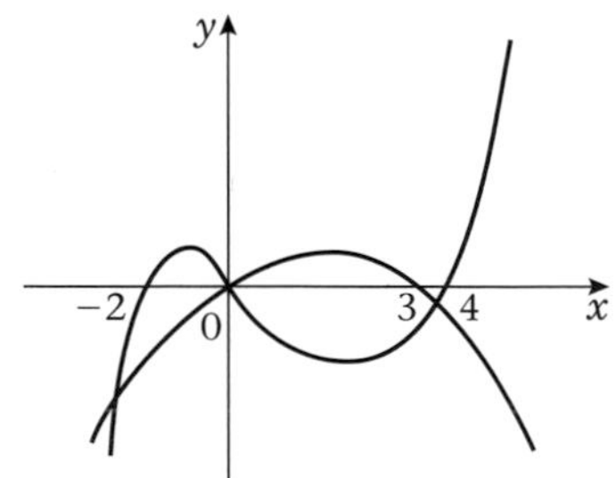

(−2, 0), (0, 0) and (4, 0)
(0, 0) and (3, 0)

b (0, 0),
$(\frac{1}{2}(1 + 3\sqrt{5}), -10 + 3\sqrt{5})$,
$(\frac{1}{2}(1 - 3\sqrt{5}), -10 - 3\sqrt{5})$

Exercise 5A

1 a -2 **b** -1 **c** 3 **d** $\frac{1}{3}$
e $-\frac{2}{3}$ **f** $\frac{5}{4}$ **g** $\frac{1}{2}$ **h** 2
i $\frac{1}{2}$ **j** $\frac{1}{2}$ **k** -2 **l** $-\frac{3}{2}$

2 a 4 **b** -5 **c** $-\frac{2}{3}$ **d** 0
e $\frac{7}{5}$ **f** 2 **g** 2 **h** -2
i 9 **j** -3 **k** $\frac{3}{2}$ **l** $-\frac{1}{2}$

3 a $4x - y + 3 = 0$ **b** $3x - y - 2 = 0$
c $6x + y - 7 = 0$ **d** $4x - 5y - 30 = 0$
e $5x - 3y + 6 = 0$ **f** $7x - 3y = 0$
g $14x - 7y - 4 = 0$ **h** $27x + 9y - 2 = 0$
i $18x + 3y + 2 = 0$ **j** $2x + 6y - 3 = 0$
k $4x - 6y + 5 = 0$ **l** $6x - 10y + 5 = 0$

4 $y = 5x + 3$

5 $2x + 5y + 20 = 0$

6 $y = -\frac{1}{2}x + 7$

7 $y = \frac{2}{3}x$

8 $(3, 0)$

9 $(\frac{5}{3}, 0)$

10 $(0, 5)$, $(-4, 0)$

Exercise 5B

1 a $\frac{1}{2}$ **b** $\frac{1}{6}$ **c** $-\frac{3}{5}$ **d** 2
e -1 **f** $\frac{1}{2}$ **g** $\frac{1}{2}$ **h** 8
i $\frac{2}{3}$ **j** -4 **k** $-\frac{1}{3}$ **l** $-\frac{1}{2}$
m 1 **n** $\dfrac{q^2 - p^2}{q - p} = q + p$

2 7

3 12

4 $4\frac{1}{3}$

5 $2\frac{1}{4}$

6 $\frac{1}{4}$

7 26

8 -5

Exercise 5C

1 a $y = 2x + 1$ **b** $y = 3x + 7$
c $y = -x - 3$ **d** $y = -4x - 11$
e $y = \frac{1}{2}x + 12$ **f** $y = -\frac{2}{3}x - 5$
g $y = 2x$ **h** $y = -\frac{1}{2}x + 2b$

2 $y = 3x - 6$

3 $y = 2x + 8$

4 $2x - 3y + 24 = 0$

5 $-\frac{1}{5}$

6 $y = \frac{2}{5}x + 3$

7 $2x + 3y - 12 = 0$

8 $\frac{8}{5}$

9 $y = \frac{4}{3}x - 4$

10 $6x + 15y - 10 = 0$

Exercise 5D

1 a $y = 4x - 4$ **b** $y = x + 2$
c $y = 2x + 4$ **d** $y = 4x - 23$
e $y = x - 4$ **f** $y = \frac{1}{2}x + 1$
g $y = -4x - 9$ **h** $y = -8x - 33$
i $y = \frac{6}{5}x$ **j** $y = \frac{2}{7}x + \frac{5}{14}$

2 $(-3, 0)$

3 $(0, 1)$

4 $(0, 3\frac{1}{2})$

5 $y = -\frac{4}{5}x + 4$

6 $x - y + 5 = 0$

7 $y = -\frac{3}{8}x + \frac{1}{2}$

8 $y = 4x + 13$

9 $y = x + 2$, $y = -\frac{1}{6}x - \frac{1}{3}$, $y = -6x + 23$

10 $(3, -1)$

Exercise 5E

1 a Perpendicular **b** Parallel
c Neither **d** Perpendicular
e Perpendicular **f** Parallel
g Parallel **h** Perpendicular
i Perpendicular **j** Parallel
k Neither **l** Perpendicular

2 $y = -\frac{1}{3}x$

3 $4x - y + 15 = 0$

4 a $y = -2x + \frac{1}{2}$ **b** $y = \frac{1}{2}x$
c $y = -x - 3$ **d** $y = \frac{1}{2}x - 8$

5 a $y = 3x + 11$ **b** $y = -\frac{1}{3}x + \frac{13}{3}$
c $y = \frac{2}{3}x + 2$ **d** $y = -\frac{3}{2}x + \frac{17}{2}$

6 $3x + 2y - 5 = 0$

7 $7x - 4y + 2 = 0$

Mixed exercise 5F

1 a $y = -3x + 14$ **b** $(0, 14)$

2 a $y = -\frac{1}{2}x + 4$ **b** $y = -\frac{1}{2}x + \frac{3}{2}$, $(1, 1)$

3 a $y = \frac{1}{7}x + \frac{12}{7}$, $y = -x + 12$ **b** $(9, 3)$

4 a $y = -\frac{5}{12}x + \frac{11}{6}$ **b** -22

5 a $y = \frac{3}{2}x - \frac{3}{2}$ **b** $(3, 3)$

6 $11x - 10y + 19 = 0$

7 a $y = -\frac{1}{2}x + 3$ **b** $y = \frac{1}{4}x + \frac{9}{4}$

8 a $y = \frac{3}{2}x - 2$ **b** $(4, 4)$ **c** 20

9 a $2x + y = 20$ **b** $y = \frac{1}{3}x + \frac{4}{3}$

10 a $\frac{1}{2}$ **b** 6 **c** $2x + y - 16 = 0$

11 a $\left(\dfrac{3 + \sqrt{3}}{1 + \sqrt{3}}\right) = \sqrt{3}$ **b** $y = \sqrt{3}x + 2\sqrt{3}$

12 a $7x + 5y - 18 = 0$ **b** $\frac{162}{35}$

13 b $y = \frac{1}{3}x + \frac{1}{3}$

14 a

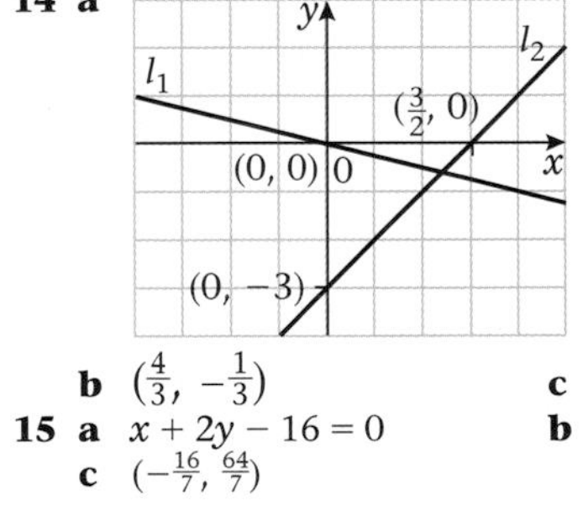

b $(\frac{4}{3}, -\frac{1}{3})$ **c** $12x - 3y - 17 = 0$

15 a $x + 2y - 16 = 0$ **b** $y = -4x$
c $(-\frac{16}{7}, \frac{64}{7})$

Exercise 6A

1 24, 29, 34
Add 5 to previous term
2 2, −2, 2
Multiply previous term by −1
3 18, 15, 12
Subtract 3 from previous term
4 162, 486, 1458
Multiply previous term by 3
5 $\frac{1}{4}, -\frac{1}{8}, +\frac{1}{16}$
Multiply previous term by $-\frac{1}{2}$
6 41, 122, 365
Multiply previous term by 3 then −1
7 8, 13, 21
Add together the two previous terms
8 $\frac{5}{9}, \frac{6}{11}, \frac{7}{13}$
Add 1 to previous numerator, add 2 to previous denominator
9 2.0625, 2.031 25, 2.015 625
Divide previous term by 2 then +1
10 24, 35, 48
Add consecutive odd numbers to previous term

Exercise 6B

1
a	$U_1 = 5$	$U_2 = 8$	$U_3 = 11$	$U_{10} = 32$
b	$U_1 = 7$	$U_2 = 4$	$U_3 = 1$	$U_{10} = -20$
c	$U_1 = 6$	$U_2 = 9$	$U_3 = 14$	$U_{10} = 105$
d	$U_1 = 4$	$U_2 = 1$	$U_3 = 0$	$U_{10} = 49$
e	$U_1 = -2$	$U_2 = 4$	$U_3 = -8$	$U_{10} = 1024$
f	$U_1 = \frac{1}{3}$	$U_2 = \frac{1}{2}$	$U_3 = \frac{3}{5}$	$U_{10} = \frac{5}{6}$
g	$U_1 = -\frac{1}{3}$	$U_2 = \frac{1}{2}$	$U_3 = -\frac{3}{5}$	$U_{10} = \frac{5}{6}$
h	$U_1 = -1$	$U_2 = 0$	$U_3 = 1$	$U_{10} = 512$

2 **a** 14 **b** 9 **c** 11 **d** 9
e 6 **f** 9 **g** 8 **h** 14
i 4 **j** 5
3 $U_n = 4n^2 + 4n = 4(n^2 + n)$ which is a multiple of 4
4 $U_n = (n - 5)^2 + 2 > 0$ U_n is smallest when $n = 5$ (U_n = 2)
5 $a = 12, b = -22$
6 $a = 1, b = 3, c = 0$
7 $p = \frac{1}{2}, q = 5\frac{1}{2}$

Exercise 6C

1 **a** 1, 4, 7, 10 **b** 9, 4, −1, −6
c 3, 6, 12, 24 **d** 2, 5, 11, 23
e 10, 5, 2.5, 1.25 **f** 2, 3, 8, 63
g 3, 5, 13, 31
2 **a** $U_{k+1} = U_k + 2, U_1 = 3$
b $U_{k+1} = U_k - 3, U_1 = 20$
c $U_{k+1} = 2U_k, U_1 = 1$
d $U_{k+1} = \frac{U_k}{4}, U_1 = 100$
e $U_{k+1} = -1 \times U_k, U_1 = 1$
f $U_{k+1} = 2U_k + 1, U_1 = 3$
g $U_{k+1} = (U_k)^2 + 1, U_1 = 0$
h $U_{k+1} = \frac{U_k + 2}{2}, U_1 = 26$
i $U_{k+2} = U_{k+1} + U_k, U_1 = 1, U_2 = 1$
j $U_{k+1} = 2U_k + 2(-1)^{k+1}, U_1 = 4$
3 **a** $U_{k+1} = U_k + 2, U_1 = 1$
b $U_{k+1} = U_k + 3, U_1 = 5$
c $U_{k+1} = U_k + 1, U_1 = 3$
d $U_{k+1} = U_k + \frac{1}{2}, U_1 = 1$
e $U_{k+1} = U_k + 2k + 1, U_1 = 1$
f $U_{k+1} = U_k - (-1)^k(2k + 1), U_1 = -1$
4 **a** $3k + 2$ **b** $3k^2 + 2k + 2$ **c** $\frac{10}{3}, -4$
5 **a** $4 - 2p$ **b** $4 - 6p$ **c** $p = -2$

Exercise 6D

1 Arithmetic sequences are **a, b, c, h, l**
2 **a** 23, $2n + 3$ **b** 32, $3n + 2$
c −3, $27 - 3n$ **d** 35, $4n - 5$
e $10x, nx$ **f** $a + 9d, a + (n - 1)d$
3 **a** £5800 **b** £(3800 + 200m)
4 **a** 22 **b** 40 **c** 39
d 46 **e** 18 **f** n

Exercise 6E

1 **a** 78, $4n - 2$ **b** 42, $2n + 2$
c 23, $83 - 3n$ **d** 39, $2n - 1$
e −27, $33 - 3n$ **f** 59, $3n - 1$
g $39p, (2n - 1)p$ **h** $-71x, (9 - 4n)x$
2 **a** 30 **b** 29 **c** 32
d 31 **e** 221 **f** 77
3 $d = 6$
4 $a = 36, d = -3$, 14th term
5 24
6 $x = 5$; 25, 20, 15
7 $x = \frac{1}{2}, x = 8$

Exercise 6F

1 **a** 820 **b** 450 **c** −1140
d −294 **e** 1440 **f** 1425
g −1155 **h** $21(11x + 1)$
2 **a** 20 **b** 25
c 65 **d** 4 or 14 (2 answers)
3 2550
4 **i** £222 500 **ii** £347 500
5 1683, 3267
6 £9.03, 141 days
7 $d = -\frac{1}{2}, -5.5$
8 $a = 6, d = -2$

Exercise 6G

1 **a** $\sum_{r=1}^{10}(3r + 1)$ **b** $\sum_{r=1}^{30}(3r - 1)$
c $\sum_{r=1}^{11}4(11 - r)$ **d** $\sum_{r=1}^{16}6r$
2 **a** 45 **b** 210
c 1010 **d** 112
3 19
4 49

Mixed exercise 6H

1 5, 8, 11
2 10
3 2, 9, 23, 51
4 **a** Add 6 to the previous term, i.e. $U_{n+1} = U_n + 6$ (or $U_n = 6n - 1$)
b Add 3 to the previous term, i.e. $U_{n+1} = U_n + 3$ (or $U_n = 3n$)
c Multiply the previous term by 3, i.e. $U_{n+1} = 3U_n$ (or $U_n = 3^{n-1}$)
d Subtract 5 from the previous term, i.e. $U_{n+1} = U_n - 5$ (or $U_n = 15 - 5n$)
e The square numbers ($U_n = n^2$)
f Multiply the previous term by 1.2, i.e. $U_{n+1} = 1.2U_n$ (or $U_n = (1.2)^{n-1}$)

Arithmetic sequences are:

a $a = 5$, $d = 6$
b $a = 3$, $d = 3$
d $a = 10$, $d = -5$

5 **a** 81 **b** 860
6 **b** 5050
7 32
8 **a** £13 780
c £42 198
9 **a** $a = 25$, $d = -3$ **b** -3810
10 **a** 26 733 **b** 53 467
11 **a** 5 **b** 45
12 **a** $-4k + 15$
b $-8k^2 + 30k - 30$
c $-\frac{1}{4}$, 4
13 **b** 1500 m
15 **a** $U_2 = 2k - 4$, $U_3 = 2k^2 - 4k - 4$
b 5, -3
16 **a** £2450
b £59 000
c $d = 30$
17 **a** $d = 5$
b 59
18 **b** $\frac{11k - 9}{3}$
c 1.5
d 415

Exercise 7A

1 **a** **i** 7 **ii** 6.5 **iii** 6.1
iv 6.01 **v** $h + 6$
b 6
2 **a** **i** 9 **ii** 8.5 **iii** 8.1
iv 8.01 **v** $8 + h$
b 8

Exercise 7B

1 $7x^6$ **2** $8x^7$ **3** $4x^3$
4 $\frac{1}{3}x^{-\frac{2}{3}}$ **5** $\frac{1}{4}x^{-\frac{4}{3}}$ **6** $\frac{1}{3}x^{-\frac{2}{3}}$
7 $-3x^{-4}$ **8** $-4x^{-5}$ **9** $-2x^{-3}$
10 $-5x^{-6}$ **11** $-\frac{1}{3}x^{-\frac{4}{3}}$ **12** $-\frac{1}{2}x^{-\frac{3}{2}}$
13 $-2x^{-3}$ **14** 1 **15** $3x^2$
16 $9x^8$ **17** $5x^4$ **18** $3x^2$

Exercise 7C

1 **a** $4x - 6$ **b** $x + 12$ **c** $8x$
d $16x + 7$ **e** $4 - 10x$
2 **a** 12 **b** 6 **c** 7
d $2\frac{1}{2}$ **e** -2 **f** 4
3 4, 0
4 $(-1, -8)$
5 1, -1
6 6, -4

Exercise 7D

1 **a** $4x^3 - x^{-2}$ **b** $-x^{-3}$ **c** $-x^{-\frac{3}{2}}$
2 **a** 0 **b** $11\frac{1}{2}$
3 **a** $(2\frac{1}{2}, -6\frac{1}{4})$ **b** $(4, -4)$ and $(2, 0)$
c $(16, -31)$ **d** $(\frac{1}{2}, 4)$ $(-\frac{1}{2}, -4)$

Exercise 7E

1 **a** $x^{-\frac{1}{2}}$ **b** $-6x^{-3}$ **c** $-x^{-4}$
d $\frac{4}{3}x^3 - 2x^2$ **e** $-6x^{-4} + \frac{1}{2}x^{-\frac{1}{2}}$
f $\frac{1}{3}x^{-\frac{2}{3}} - \frac{1}{2}x^{-2}$ **g** $-3x^{-2}$ **h** $3 + 6x^{-2}$
i $5x^{\frac{3}{2}} + \frac{3}{2}x^{-\frac{1}{2}}$ **j** $3x^2 - 2x + 2$ **k** $12x^3 + 18x^2$
l $24x - 8 + 2x^{-2}$
2 **a** 1 **b** $\frac{2}{9}$ **c** -4 **d** 4

Exercise 7F

1 $24x + 3$, 24
2 $15 - 3x^{-2}$, $6x^{-3}$
3 $\frac{9}{2}x^{-\frac{1}{2}} + 6x^{-3}$, $-\frac{9}{4}x^{-\frac{3}{2}} - 18x^{-4}$
4 $30x + 2$, 30
5 $-3x^{-2} - 16x^{-3}$, $6x^{-3} + 48x^{-4}$

Exercise 7G

1 $2t - 3$ **2** 2π
3 $-\frac{4}{3}$ **4** 151
5 18

Exercise 7H

1 **a** $y + 3x - 6 = 0$ **b** $4y - 3x - 4 = 0$
c $3y - 2x - 18 = 0$ **d** $y = x$
e $y = 12x + 14$ **f** $y = 16x - 22$
2 **a** $7y + x - 48 = 0$ **b** $17y + 2x - 212 = 0$
3 $(1\frac{2}{9}, 1\frac{8}{9})$
4 $y = -x$, $4y + x - 9 = 0$; $(-3, 3)$
5 $y = -8x + 10$, $8y - x - 145 = 0$

Exercise 7I

1 4, $11\frac{3}{4}$, $17\frac{25}{27}$
2 0, $\pm 2\sqrt{2}$
3 $(-1, 0)$ and $(1\frac{2}{3}, 9\frac{13}{27})$
4 2, $2\frac{2}{3}$
5 $(2, -13)$ and $(-2, 15)$
6 **a** $1 - \dfrac{9}{x^2}$ **b** $x = \pm 3$
7 $x = 4$, $y = 20$
8 $\frac{3}{2}x^{-\frac{1}{2}} + 2x^{-\frac{3}{2}}$
9 **a** $\dfrac{dy}{dx} = 6x^{-\frac{1}{2}} - \frac{3}{2}x^{\frac{1}{2}}$ **b** $(4, 16)$
$= \frac{1}{2}x^{-\frac{1}{2}}(12 - 3x)$
$= \frac{3}{2}x^{-\frac{1}{2}}(4 - x)$
10 **a** $x + x^{\frac{3}{2}} - x^{-\frac{1}{2}} - 1$
b $1 + \frac{3}{2}x^{\frac{1}{2}} + \frac{1}{2}x^{-\frac{3}{2}}$
c $4\frac{1}{16}$
11 $6x^2 + \frac{1}{2}x^{-\frac{1}{2}} - 2x^{-2}$
12 $\dfrac{10}{3}, \dfrac{2300\pi}{27}$
14 $a = 1$, $b = -4$, $c = 5$
15 **a** $3x^2 - 10x + 5$
b **i** $\frac{1}{3}$ **ii** $y = 2x - 7$ **iii** $\frac{7}{2}\sqrt{5}$
16 $y = 9x - 4$ and $9y + x = 128$
17 **a** $(\frac{4}{5}, -\frac{2}{5})$ **b** $\frac{1}{5}$

Exercise 8A

1 $y = \frac{1}{6}x^6 + c$
2 $y = 2x^5 + c$
3 $y = x^3 + c$
4 $y = x^{-1} + c$
5 $y = 2x^{-2} + c$
6 $y = \frac{3}{5}x^{\frac{5}{3}} + c$
7 $y = \frac{8}{3}x^{\frac{3}{2}} + c$
8 $y = -\frac{2}{7}x^7 + c$
9 $y = \frac{1}{2}x^6 + c$
10 $y = -x^{-3} + c$
11 $y = 2x^{\frac{1}{2}} + c$
12 $y = -10x^{-\frac{1}{2}} + c$
13 $y = 4x^{-\frac{1}{2}} + c$
14 $y = \frac{9}{2}x^{\frac{4}{3}} + c$
15 $y = 3x^{12} + c$
16 $y = 2x^{-7} + c$
17 $y = -9x^{\frac{1}{3}} + c$
18 $y = -5x + c$
19 $y = 3x^2 + c$
20 $y = \frac{10}{3}x^{0.6} + c$

Exercise 8B

1 a $y = 2x^2 + x^{-1} + 4x^{\frac{3}{2}} + c$
b $y = 5x^3 - 3x^{-2} + 2x^{-\frac{3}{2}} + c$
c $y = \frac{1}{4}x^4 - 3x^{\frac{1}{2}} + 6x^{-1} + c$
d $y = x^4 + 3x^{\frac{1}{3}} + x^{-1} + c$
e $y = 4x + 4x^{-3} + 4x^{\frac{1}{2}} + c$
f $y = 3x^{\frac{5}{3}} - 2x^5 - \frac{1}{2}x^{-2} + c$
g $y = 4x^{-\frac{1}{3}} - 3x + 4x^2 + c$
h $y = x^5 + 2x^{-\frac{1}{2}} + 3x^{-4} + c$

2 a $f(x) = 6x^2 - 3x^{-\frac{1}{2}} + 5x + c$
b $f(x) = x^6 - x^{-6} + x^{-\frac{1}{6}} + c$
c $f(x) = x^{\frac{1}{2}} + x^{-\frac{1}{2}} + c$
d $f(x) = 5x^2 - 4x^{-2} + c$
e $f(x) = 3x^{\frac{2}{3}} - 6x^{-\frac{2}{3}} + c$
f $f(x) = 3x^3 - 2x^{-2} + \frac{1}{2}x^{\frac{1}{2}} + c$
g $f(x) = \frac{1}{3}x^3 - x^{-1} + \frac{2}{3}x^{\frac{3}{2}} + c$
h $f(x) = x^{-2} - x^2 + \frac{4}{3}x^{\frac{3}{2}} + c$

Exercise 8C

1 $\frac{1}{4}x^4 + x^2 + c$
2 $-2x^{-1} + 3x + c$
3 $2x^{\frac{5}{2}} - x^3 + c$
4 $\frac{4}{3}x^{\frac{3}{2}} - 4x^{\frac{1}{2}} + 4x + c$
5 $x^4 + x^{-3} + rx + c$
6 $t^3 + t^{-1} + c$
7 $\frac{2}{3}t^3 + 6t^{-\frac{1}{2}} + t + c$
8 $\frac{1}{2}x^2 + 2x^{\frac{1}{2}} - 2x^{-\frac{1}{2}} + c$
9 $\frac{p}{5}x^5 + 2tx - 3x^{-1} + c$
10 $\frac{p}{4}t^4 + q^2t + px^3t + c$

Exercise 8D

1 a $\frac{1}{2}x^4 + x^3 + c$
b $2x - \frac{3}{x} + c$
c $\frac{4}{3}x^3 + 6x^2 + 9x + c$
d $\frac{2}{3}x^3 + \frac{1}{2}x^2 - 3x + c$
e $\frac{4}{5}x^{\frac{5}{2}} + 2x^{\frac{3}{2}} + c$

2 a $\frac{1}{3}x^3 + 2x^2 + 4x + c$
b $\frac{1}{3}x^3 + 2x - \frac{1}{x} + c$
c $\frac{1}{2}x^2 + \frac{8}{3}x^{\frac{3}{2}} + 4x + c$
d $\frac{2}{5}x^{\frac{5}{2}} + \frac{4}{3}x^{\frac{3}{2}} + c$
e $\frac{2}{3}x^{\frac{3}{2}} + 4x^{\frac{1}{2}} + c$
f $2x^{\frac{1}{2}} + \frac{4}{3}x^{\frac{3}{2}} + c$

3 a $2x^{\frac{3}{2}} - \frac{1}{x} + c$
b $4x^{\frac{1}{2}} + x^3 + c$
c $\frac{3}{5}x^{\frac{5}{3}} - \frac{2}{x^2} + c$
d $-\frac{1}{x^2} - \frac{1}{x} + 3x + c$
e $\frac{1}{4}x^4 - \frac{1}{3}x^3 + \frac{3}{2}x^2 - 3x + c$
f $4x^{\frac{1}{2}} + \frac{6}{5}x^{\frac{5}{2}} + c$
g $\frac{1}{3}x^3 - 3x^2 + 9x + c$
h $\frac{8}{5}x^{\frac{5}{2}} + \frac{8}{3}x^{\frac{3}{2}} + 2x^{\frac{1}{2}} + c$
i $3x + 2x^{\frac{1}{2}} + 2x^3 + c$
j $\frac{2}{5}x^{\frac{5}{2}} + 3x^2 + 6x^{\frac{3}{2}} + c$

Exercise 8E

1 a $y = x^3 + x^2 - 2$
b $y = x^4 - \frac{1}{x^2} + 3x + 1$
c $y = \frac{2}{3}x^{\frac{3}{2}} + \frac{1}{12}x^3 + \frac{1}{3}$
d $y = 6\sqrt{x} - \frac{1}{2}x^2 - 4$
e $y = \frac{1}{3}x^3 + 2x^2 + 4x + \frac{2}{3}$
f $y = \frac{2}{5}x^{\frac{5}{2}} + 6x^{\frac{1}{2}} + 1$

2 $f(x) = \frac{1}{2}x^4 + \frac{1}{x} + \frac{1}{2}$

3 $y = 1 - \frac{2}{\sqrt{x}} - \frac{3}{x}$

4 a $f_2(x) = \frac{x^3}{3}$; $f_3(x) = \frac{x^4}{12}$
b $\frac{x^{n+1}}{3 \times 4 \times 5 \times \ldots \times (n+1)}$

5 $f_2(x) = x + 1$; $f_3(x) = \frac{1}{2}x^2 + x + 1$;
$f_4(x) = \frac{1}{6}x^3 + \frac{1}{2}x^2 + x + 1$

Mixed Exercise 8F

1 a $\frac{2}{3}x^3 - \frac{3}{2}x^2 - 5x + c$
b $\frac{3}{4}x^{\frac{4}{3}} + \frac{3}{2}x^{\frac{2}{3}} + c$

2 $\frac{1}{3}x^3 - \frac{3}{2}x^2 + \frac{2}{x} + \frac{1}{6}$

3 a $2x^4 - 2x^3 + 5x + c$
b $2x^{\frac{5}{2}} + \frac{4}{3}x^{\frac{3}{2}} + c$

4 $\frac{4}{5}x^{\frac{5}{2}} - \frac{2}{3}x^{\frac{3}{2}} - 6x^{\frac{1}{2}} + c$

5 $x = t^3 - t^2 + t + 1$; $x = 7$

6 $2x^{\frac{3}{2}} + 4x^{\frac{1}{2}} + c$

7 $x = 12\frac{1}{3}$

8 a $A = 6$. $B = 9$
b $\frac{3}{5}x^{\frac{5}{3}} + \frac{9}{2}x^{\frac{4}{3}} + 9x + c$

9 a $\frac{3}{2}x^{-\frac{1}{2}} + 2x^{-\frac{3}{2}}$
b $2x^{\frac{3}{2}} - 8x^{\frac{1}{2}} + c$

10 a $5x - 8x^{\frac{1}{2}} - \frac{2}{3}x^{\frac{3}{2}} + c$

Review Exercise 2 (Chapters 5 to 8)

1 a Since P(3, −1), substitute values into $y = 5 - 2x$ gives $-1 = 5 - 6$, so P on line.
b $x - 2y - 5 = 0$
2 a $AB = 5\sqrt{2}$
b $0 = x - 7y + 9$
c C is $(0, \frac{9}{7})$
3 a $0 = x - 3y - 21$
b $P = (3, -6)$
c 10.5 units^2
4 a $p = 15$, $q = -3$
b $7x - 5y - 46 = 0$
c $x = 11\frac{4}{7}$
5 a $y = -\frac{1}{3}x + 4$
b C is (3, 3)
c 15 units^2
6 a P is $(\frac{11}{8}, \frac{13}{16})$
b $\frac{121}{64}$ units^2
7 a $d = 3.5$
b $a = -10$
c 217.5
8 $a = 5$ km, $d = 0.4$ km
9 a −3, −1, 1
b $d = 2$
c $n(n - 4)$
10 a £750
b £14 500
c £155
11 a $a_2 = 4$, $a_3 = 7$
b 73
12 a $a_1 = k$, $a_2 = 3k + 5$
b $a_3 = 3a_2 + 5 = 9k + 20$
c i $40k + 90$ **ii** $10(4k + 9)$
13 a $a_5 = 16k - 45$
b $k = 4$
c 81
14 a In general:

$$S_n = a + (a + d) + (a + 2d) + \ldots + (a + (n - 2)d) + (a + (n - 1)d)$$

Reversing the sum:

$$S_n = (a + (n - 1)d) + (a + (n - 2)d) + (a + (n - 3)d) + \ldots + (a + d) + a$$

Adding the two sums:

$$2S_n = [2a + (n - 1)d] + [2a + (n - 1)d] + \ldots + [2a + (n - 1)d]$$

$$2S_n = n[2a + (n - 1)d]$$

$$S_n = \frac{n}{2}[2a + (n - 1)d]$$

b £109
c $n^2 - 150n + 5000 = 0$
d $n = 50$ or 100
e $n = 100$ (gives a negative repayment)
15 a $a_2 = 4 - 2k$
$a_3 = (4 - 2k)^2 - k(4 - 2k) = 6k^2 - 20k + 16$
b $k = 1$ or $k = \frac{7}{3}$
c $a_2 = -\frac{2}{3}$
d $a_5 = 2$
e $a_{100} = -\frac{2}{3}$
16 $\frac{dy}{dx} = 12x^2 + x^{-\frac{1}{2}}$
17 a $\frac{dy}{dx} = 4x + 18x^{-4}$
b $\frac{2}{3}x^3 + 3x^{-2} + c$
18 a $\frac{dy}{dx} = 6x + 2x^{-\frac{1}{2}}$
b $\frac{d^2y}{dx^2} = 6 - x^{-\frac{3}{2}}$
c $x^3 + \frac{8}{3}x^{\frac{3}{2}} + c$
19 a i $\frac{dy}{dx} = 15x^2 + 7$ **ii** $\frac{d^2y}{dx^2} = 30x$
b $x + 2x^{\frac{3}{2}} + x^{-1} + c$
20 a $\frac{dy}{dx} = 4 + \frac{9}{2}x^{\frac{1}{2}} - 4x$
b Substitute values, $8 = 8$
c $3y = x + 20$
d $PQ = 8\sqrt{10}$
21 a $\frac{dy}{dx} = 8x - 5x^{-2}$, at P this is 3
b $y = 3x + 5$
c $k = -\frac{5}{3}$
22 a At (3, 0), $y = 0$
b At P, $y = -7x + 21$
c $Q = (5, -15\frac{1}{3})$
23 a $P = 2$, $Q = 9$, $R = 4$
b $3x^{\frac{1}{2}} + \frac{9}{2}x^{-\frac{1}{2}} - 2x^{-\frac{3}{2}}$
c When $x = 1$, $f'(x) = 5\frac{1}{2}$, gradient of $2y = 11x + 3$ is $5\frac{1}{2}$, so it is parallel with tangent.
24 $\frac{1}{3}x^3 + 2x^2 - 3x - 13$
25 $3x + 2x^{\frac{5}{2}} + 4x^{\frac{1}{2}} - 3$
26 a $3x^2 + 2$
b $3x^2 + 2 \geqslant 2$ for all values of x since $3x^2 \geqslant 0$ for all values of x
c $y = \frac{1}{4}x^4 + x^2 - 7x + 10$
d $5y + x - 22 = 0$
27 a $y = -\frac{1}{3}x^{-3} + x^{-1} + \frac{4}{3}$
b (1, 2) and $(-1, \frac{2}{3})$
28 a $2x^3 - 5x^2 - 12x$
b $x(2x + 3)(x - 4)$
c

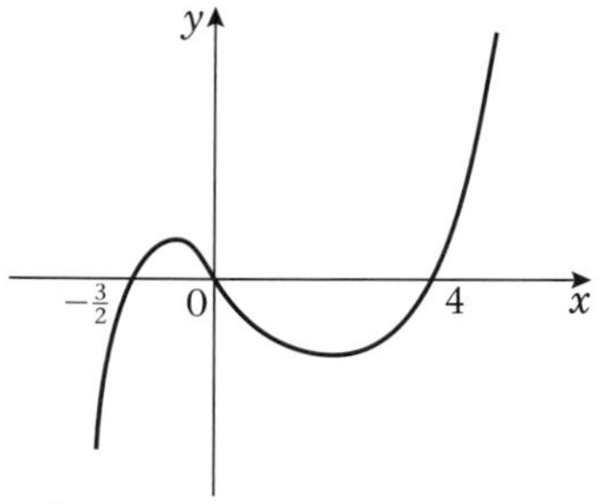

$(-\frac{3}{2}, 0)$, (0, 0) and (4, 0)
29 a $PQ^2 = 1^2 + 13^2 = 170$
$PQ = \sqrt{170}$
b $\frac{dy}{dx} = 3x^2 - 12x - 4x^{-2}$
At P, $\frac{dy}{dx} = -13$, at Q, $\frac{dy}{dx} = -13$
c $x - 13y - 14 = 0$
30 a $x(x - 3)(x - 4)$

b 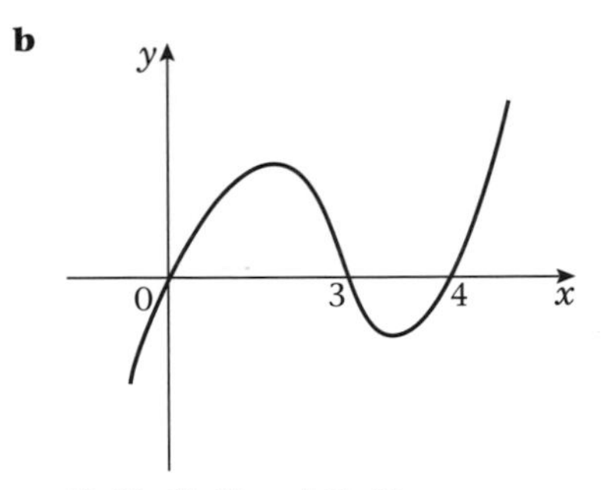

(0, 0), (3, 0) and (4, 0)

c P = $(3\frac{4}{7}, -1\frac{5}{7})$

Practice paper

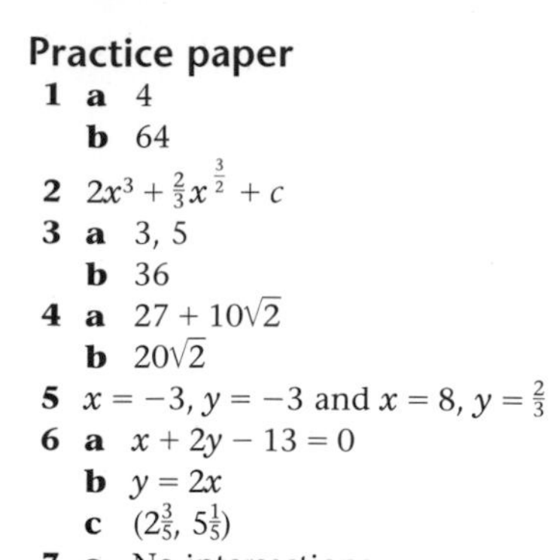

1 **a** 4

b 64

2 $2x^3 + \frac{2}{3}x^{\frac{3}{2}} + c$

3 **a** 3, 5

b 36

4 **a** $27 + 10\sqrt{2}$

b $20\sqrt{2}$

5 $x = -3, y = -3$ and $x = 8, y = \frac{2}{3}$

6 **a** $x + 2y - 13 = 0$

b $y = 2x$

c $(2\frac{3}{5}, 5\frac{1}{5})$

7 **a** No intersections.

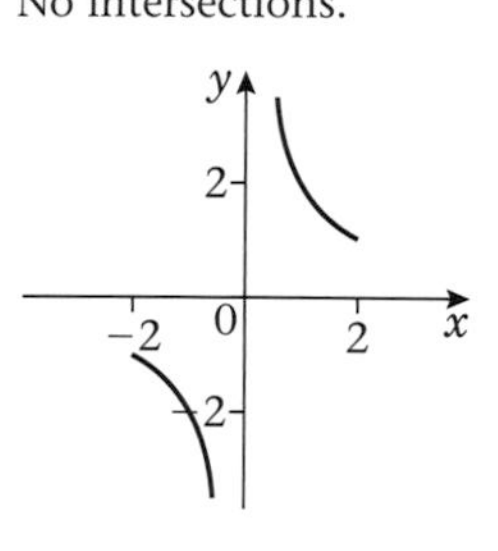

b

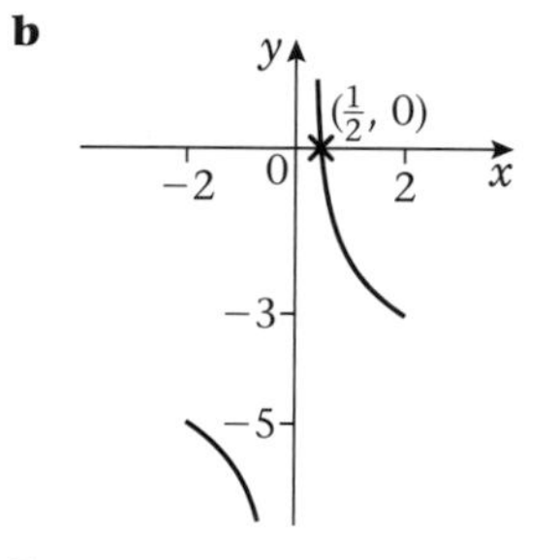

c

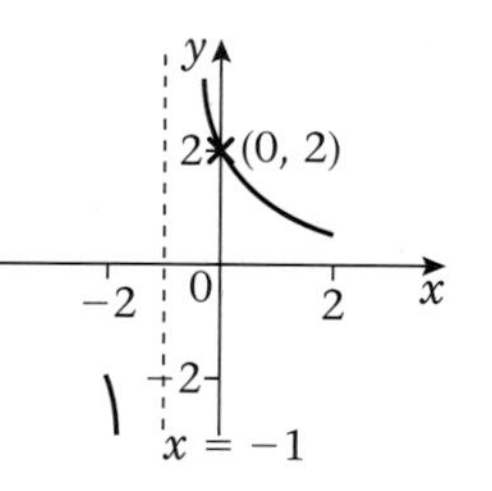

8 **a** 670 **b** 5350 **c** 45

9 **a** **i** 2 **ii** $c - 4$ **iii** $c < 4$

b **i** $x < 5$

ii $x < -7, x > 3$

iii $x < -7, 3 < x < 5$

10 **a** $P = 9, Q = -24, R = 16$

b 10

c $x + 10y - 248 = 0$

Examination style paper

1 a $k = 5$
b $k = 6$
2 a 9
b $8x^{-\frac{1}{3}}$
3 a

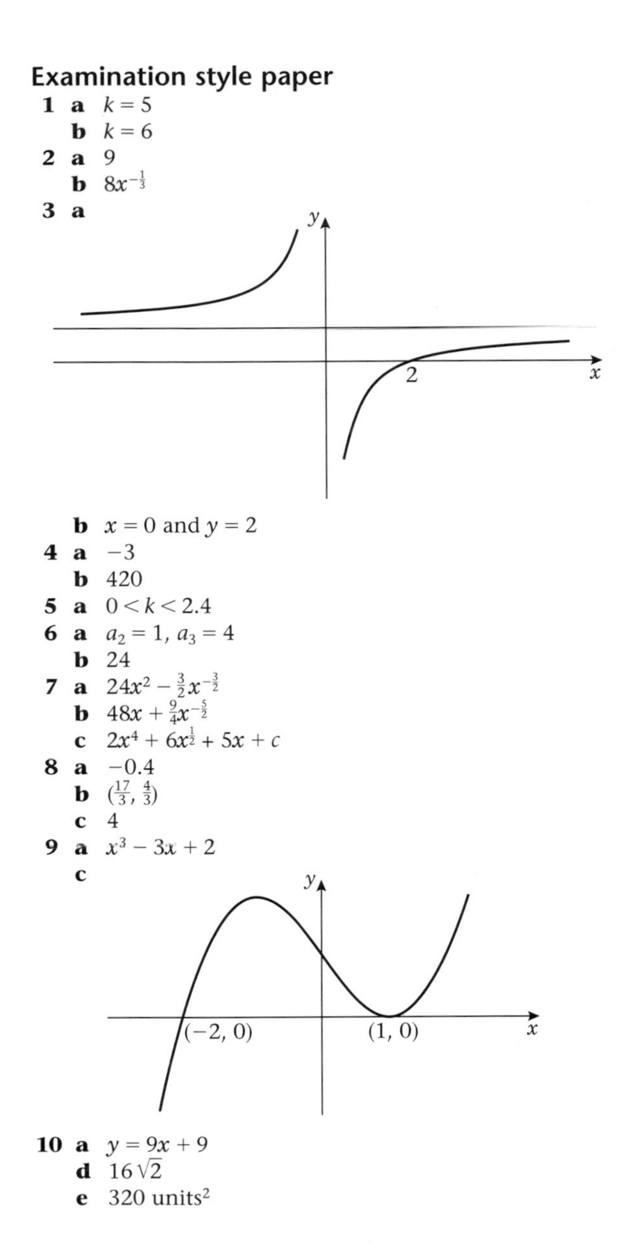

b $x = 0$ and $y = 2$
4 a -3
b 420
5 a $0 < k < 2.4$
6 a $a_2 = 1$, $a_3 = 4$
b 24
7 a $24x^2 - \frac{3}{2}x^{-\frac{3}{2}}$
b $48x + \frac{9}{4}x^{-\frac{5}{2}}$
c $2x^4 + 6x^{\frac{1}{2}} + 5x + c$
8 a -0.4
b $(\frac{17}{3}, \frac{4}{3})$
c 4
9 a $x^3 - 3x + 2$
c

10 a $y = 9x + 9$
d $16\sqrt{2}$
e 320 units^2

Index